Moths to the Flame

Moths to the Flame
The Seductions of Computer Technology

Gregory J. E. Rawlins

The empires of the future are the empires of the mind.
Winston Churchill, *Onwards to Victory*

A Bradford Book
The MIT Press
Cambridge, Massachusetts
London, England

This book was set in Sabon by Windfall Software using ZzTEX and was printed and bound in the United States of America.

Library of Congress Cataloging-in-Publication Data
Rawlins, Gregory J. E.
   Moths to the flame : the seductions of computer technology / Gregory J. E.
Rawlins
     p. cm.
   Includes index.
   ISBN 0-262-18176-2 (hard : alk. paper)
   1. Computers and civilization.   I. Title.
QA76.9.C66R39   1996
303.48'34—dc20   96-4729
                                                 CIP

To the next generation,
Both human and machine.

Yet it is not our part to master all the tides of the world, but to do what is in us for the succour of those years wherein we are set, uprooting the evil in the fields that we know, so that those who live after may have clean earth to till. What weather they shall have is not ours to rule.

J. R. R. Tolkien, *The Lord of the Rings*

To the next generation,
Both human and machine.

Yet it is not our part to master all the tides of the world, but to do what is in us for the succour of those years wherein we are set, uprooting the evil in the fields that we know, so that those who live after may have clean earth to till. What weather they shall have is not ours to rule.

J. R. R. Tolkien, *The Lord of the Rings*

Contents

Preface

Any philosophy that can be put in a nutshell belongs there.
Sidney J. Harris, *Leaving the Surface*

I've always been interested in the future. As they say, the future is where I will spend the rest of my life. When I was a kid, that future meant rocket ships and food pills; faster transportation, space exploration, and miracle medication; docile robots, clean air, and no violence.

Well, we're in that future now. But it isn't much like the future we foresaw way back when. Today, the new thing is the computer. Instead of flying cars or household robots, most of us are getting a foretaste of the future by monkeying around with the expensive paperweights on our desks. What are they capable of? How will they change? How will we change? No one knows.

I go to the supermarket and blithely hand over a credit card to pay for my groceries. Then I wonder whether this particular store is keeping track of that information. What are they doing with it? Who are they selling it to?

I go to the movies and marvel at the special effects. Then wonder how it was all done. Then I think about how the technology that will make future movies possible might be used in other ways. Future propaganda might compare to today's persuasions the way a child's crayon drawing compares to the Sistine Chapel.

I meet friends who treat the choice of their portable computer as a fashion statement—like the clothes they wear or the car they drive. What will happen when computers grow even smaller and more powerful? How will we use them as status symbols?

I switch on my television set and watch live satellite feeds from a made-for-TV war in the Middle East. Sitting there, stapled to the screen, I wonder where weapons technology is going—and what's going to happen to us when it gets there.

Today, concerns over privacy and worldwide computer communications have already bubbled over into public consciousness, but many deeper and longer-term issues, like future warfare and employment, have yet to do so. Many today seem to see us moving toward either a rosy, gee-whiz future where benevolent technology makes everything better, or a dark, satanic future where rampant technology tramples all our cherished institutions. Which, if any, of these two visions is right? Is there a third way?

Now that information technology is accelerating toward contributing half of the domestic product of advanced nations, it is rapidly becoming the cutting edge of social change. With so much legal, political, and economic change, who will profit and who will suffer? Who will become the information aristocracy and who will become the infoserfs of the next generation?

Our near future will be a complex, exciting, but also frightening place. It will present us with many anxious choices and unforeseen consequences. Our every decision today and for the next twenty years will represent a bet we will be placing against our better natures. Whatever the outcomes of those bets, the consequences are likely to be extreme.

This is the future we inhabit. It's one of distant voices and complex systems, of interdependent processes and new choices, of faceless strangers and wonder-making machines.

I no longer think much about the future. Because that future is now.

1

Too Many Secrets

For ye shall know the truth, and the truth shall set you free. [CIA motto]
The Bible, John 8:32

What is truth? said jesting Pilate, and would not stay for an answer.
Francis Bacon, *Essays,* "Of Truth"

Roughly five hundred years after Julius Caesar was stabbed to death by a cabal of Roman senators, the ancient Egyptians left the world a great mystery. For centuries they had written down their history using hieroglyphs, a beautiful pictorial language of signs and drawings of animals and plants. But, after long centuries of war and general disarray, knowledge of their meaning had died out. Egyptian history was now encrypted, locked in the dusty tombs and obelisks dotting the arid landscape, mute even to the modern Egyptians who moved among them, living life on the Nile as their ancestors had done for over four thousand years. The hieroglyphs weren't decrypted until some years after Napoleon's troops discovered the Rosetta stone in July of 1799.

The Rosetta stone, a slab of black basalt weighing three-quarters of a ton, now stands in the Egyptian sculpture gallery of the British Museum in London. When discovered it was probably embedded in a crumbling wall in the little village of El Rashid—called Rosetta by the British—a few kilometers from the Mediterranean on the west bank of the Nile.

The stone had been inscribed about two thousand years earlier with a priestly decree to Ptolemy V, a twelve-year-old Greek boy who was the newly crowned pharaoh of Egypt. Descended from a Greek general in the army of Alexander the Great, Ptolemy V's line continued to rule

Egypt for a further three hundred years—until Cleopatra VI had her dramatic and ultimately fatal run-in with Julius Caesar, Mark Antony, Augustus Caesar, and the rising power of Rome.

Almost two thousand years later, in the summer of 1814, an English physician and physicist named Thomas Young carried a copy of the mysterious Rosetta inscriptions with him on holiday. The inscriptions appeared to be in three different languages: ancient Egyptian hieroglyphs and two other languages, one of which was Greek, the language of the Macedonian dynasty that ended with Cleopatra.

Young's scientific bent led him to count frequently occurring Greek words, then to look for groups of hieroglyphs appearing roughly the same number of times. After four years of patient work, his statistical analysis gave him the beginnings of a dictionary of hieroglyphs to Greek words. Other scholars built on his work and a few years later completed the decryption.

Finally, we could again read the Rosetta Stone. Here's how it starts:

There being assembled the Chief Priests and Prophets and those who enter the inner shrine for the robing of the gods and the Fan-bearers and the Sacred Scribes and all the other priests from the temples throughout the land who have come to meet the king at Memphis for the feast of the assumption by PTOLEMY THE EVER-LIVING THE BELOVED OF PTAH THE GOD EPIPHANES EUCHARISTOS.

Silent for almost fourteen hundred years, the ancient Egyptians spoke again.

The Dancing Men

In Arthur Conan Doyle's "The Adventure of the Dancing Men" Doctor Watson was once again amazed at his companion's penetrating insights, this time into the cryptic messages hidden in the dancing men, a seeming child's scrawl of dancing stick figures.

When first presented with the mystery, Holmes could do nothing— a short encrypted message could mean anything: "These hieroglyphics have evidently a meaning. If it is a purely arbitrary one, it may be impossible for us to solve it. If, on the other hand, it is systematic, I have no doubt that we shall get to the bottom of it." Once presented with a few more secret messages, however, he rapidly broke the encryption, for the conspirators had foolishly used a simple encryption scheme. At the risk of spoiling the story of the dancing men, I'll give Holmes's description of how he decrypted the messages: "The first message submitted to me was so short that it was impossible for me to do more than say, with some confidence, that the symbol 𝑋 stood for E."

Clever Holmes had seen, just as Young had with the Rosetta stone, that changing each letter (or word) into another symbol may make the message look odd, but it doesn't change each letter's frequency. Moreover, like Scrabble players today, he knew that *e* is the most common letter in English. So he simply found the most frequent dancing man and assumed that it stood for *e*.

Now, in the single word I have already got the two E's coming second and fourth in a word of five letters. It might be 'sever,' or 'lever,' or 'never.' There can be no question that the latter as a reply is most probable. . . . Accepting it as correct, we are now able to say that the symbols 𝑋𝑋𝑋 stand respectively for N, V, and R.

And so, by bits and pieces, Holmes broke the encryption and unmasked the clueless conspirators.

Everyone keeps secrets. Over two thousand years ago, for example, Julius Caesar encrypted messages to his generals far afield by cyclically mapping letters to the third letter on in the alphabet: *a* became *d*, *b* became *e*, and so on, and *z* became *c*. Thus, the message "attack at dawn" would become "dwwdfn dw gdzq."

Over fifteen hundred years later, Mary, Queen of Scots, used such an encryption to plot with Spain the assassination of her cousin, Queen Elizabeth I. Sadly for Mary, Elizabeth's secret agents, who had cleverly instigated the conspiracy in the first place, used a statistical analysis to quickly break the encryption. So at eight in the morning of Wednesday, February 8, 1587, Mary lost her head.

Governments around the world took note and never again used anything as simple as a Caesar encryption. But while the new schemes they came up with were more complex, they still only substituted and rearranged letters and other symbols. Nobody could think of anything better.

So although practical secrecy advanced over the centuries—particularly during the fury of technological development we now call the Second World War—modern secrecy really only started in 1948 when Claude Shannon, a brilliant researcher at AT&T Bell Laboratories, put his finger on the real problem. Shannon saw that roughly seven in every ten letters in a long English message are redundant. For example, the three letters *e, t,* and *a* alone account for well over a quarter of all letters used in English. If English were not redundant, all letters would occur with equal frequency.

Similarly, the nine words: *the, of, and, to, it, you, be, have,* and *will,* account for a quarter of all words used. It says something about us that *I, me, my,* and *mine* didn't make the top nine, but *you* did.

Because all words and letters aren't equally likely, each letter of a word, or each word of a sentence, builds context for the next one, thereby reducing the choices. For this reason, the first few letters of a word (the first few words of a sentence, the first few sentences of a book) are usually more important (that is, less redundant) than later ones.

For example, in English *q* is always followed by *u.* Also, if a word starts with *th* the next letter is almost surely a vowel, the pseudovowel *y,* or an *r.* If a word starts with *the* chances are almost one in two that the next letter is *r.* And if a word starts with *ther* the next letter is almost surely a vowel or *m.* Similarly, if we hear the word *the* we usually expect the next word to be a noun, or the beginning of a noun phrase. If we hear the phrase *the cat ate the,* we expect the next word to be *mouse.* Of course, the next word could just as well be *television.* The more previous context there is, the fewer the possible continuations—and the greater our surprise when the next word isn't what we expected.

Languages are so redundant to help us understand garbled communication. But that massive redundancy makes long encrypted messages, even after massive rearrangements and replacements, pretty easy to break by statistical analysis. Lik■ al■ hu■an ■ang■age■, E■gli■h i■ ex■rem■ly ■edu■dan■.

Of course, even a simple encryption can take time to break; for dramatic reasons, Doyle shortened Holmes's task considerably. Besides the three possibilities he identified, the word he interpreted as *never* could have been *aedes, bedew, beget, beret,* or scores of others. Today, of course, we can find all such possibilities in millionths of a second with a computer.

Secret writers took all those lessons to heart but didn't know what to do about them. The only answer seemed to be to use their computers to pile on more and more rearrangements and replacements in greater and greater profusion, hoping that the secrecy breakers' computers weren't fast enough to keep up. But given the way computers were improving, they knew that was a losing proposition.

One If by Land, Two If by Sea

The point of having a secret is often to share it. A secret is rarely interesting if you can't tell anyone about it. Paul Revere, for example, certainly wouldn't have helped along the American Revolution if he had neglected to pass on the information about the English Redcoats he received through the lantern encryption scheme.

So let's suppose Alice and Bob have to share a secret. They may or may not trust each other; or Bob may trust Alice, while Alice doesn't trust Bob, or conversely. And whether or not they trust each other, one or both of them may mistrust the communications channel they have to use to share their secret.

For example, when meeting a friend in private, we usually trust both the friend and the channel (the private face-to-face conversation). When telephoning a friend, we probably trust the friend but not, perhaps, the channel, since someone may be eavesdropping. When meeting a stranger, we may distrust the stranger but trust the channel. And when two paranoiacs meet, they trust neither each other nor the channel.

Whenever we share information we could be sharing secrets. So whether it's cable companies sending encrypted signals to television sets or spies sending encrypted messages to governments through radio waves, smoke signals, or bongo drums, the need for secrecy is potentially all around us.

Even if Alice and Bob have an eavesdrop-proof channel, they can't use it to exchange all their secrets, because it may not be able to carry long messages or may only be usable now and again. For example, their secure channel could be a private face-to-face conversation, but they may meet infrequently, or speech may be unsuitable to pass along large amounts of information. Either way, at some point they will have to rely on a channel they know to be insecure.

Long and painful historical experience has taught us the virtue of assuming that interlopers intercept every single encrypted message and that they know everything about our security system except the actual encryption keys we use. That includes knowing all keys used previously and having translations for all earlier encrypted messages. Even then our security system should still be secure.

We give secrecy breakers so much benefit of the doubt because changing keys is easy, but changing the entire system—formats, channels, and protocols—is hard. So since those parts of the system are usually longlasting, and people are weak, it's best to assume they are all compromised.

During the Second World War the Germans failed to learn that lesson, and may have lost the war as a result. They had developed a machine to encrypt information that was broadcast by radio to their military outposts, spies, and ships and submarines at sea. What they didn't know was that a group of British decrypters had secretly built a primitive computer and, with its aid, had broken the entire system early in 1940. So often Winston Churchill sat in his office and read their secret transmissions only hours after they were broadcast.

The Germans, overconfident in their system, rarely changed keys or encryption schemes. They thought their system unbreakable and attributed all their military setbacks to an imaginary network of brilliant British spies. They placed too much faith in the secrecy of their encryption methods.

Today the rule is that in a good encryption scheme we should be able to read and reread the encryption procedure til our eyes bubble, and puzzle over previous translations til our brains hurt, and still be unable to decrypt any part of the current message—unless we also know the particular key used.

Of course, like stage magic, secrecy can rely on physical security, misdirection, and deception, as well as on encryption. For example, when the American president travels by car, the car is bulletproof, but there are also many decoy cars. The British government uses the same method to transport jailed high-profile terrorists.

Sending a secret message down a communications channel is like stashing a secret in a room. To keep it secret we might lock the room (that is, use a secure channel and keep information unencrypted); lock the information (use an insecure channel but encrypt the information); or hide the information (keep information apparently unencrypted but hide a message in it).

Of course, the truly paranoid among us, being prudent, will sensibly lock both the room and the information, hide the information, and then hide the room. Which scheme we choose depends on how paranoid we are, and how important the information's loss would be.

When our level of suspicion gets high enough, we don't even want adversaries to know that we sent a message. During the Cold War, American and British spies often bugged Soviet installations. But each time they did so they could tell that the operation had gone wrong because there was a precipitous drop in message traffic. Through several highly placed double agents in British intelligence, the Soviets always knew when someone was listening.

Going Public

Before 1975, all encryption schemes forced the sender and the receiver to have the same secret key. If Bob sends Alice an encrypted message, to read it Alice must first know Bob's key. Before 1975, all encryption schemes linked the encryption and the decryption. If you gave away your encryption key, you gave away your decryption key, because they were the same thing. Further, if you encrypted something, you could later decrypt it.

So secret-key systems are schizophrenic—they need secret, but shared, keys. And they need many keys too. If a thousand of us have to share secrets with each other using a secret-key system, we could need half a million shared keys—every pair of us using the system might have to have our own shared secret key.

Of course, we can distribute keys securely, say by armed courier, but that's expensive and slow. So the more shared secret keys we need, the more expensive the system becomes. Further, the more of us who have to know a key, the more insecure the system is and the harder it is to change keys if there's a security breach.

All in all, it was a wretched way to work, but for several thousand years no one could think of anything better. Then the computer led to something revolutionary. And because of it, it's now possible for any of us to have as much privacy as major governments have. In 1975, two computer professionals, Whitfield Diffie and Martin Hellman, invented *public-key systems,* in which everyone has two keys, one private and one public. Each person keeps the *private key* secret, but can learn anyone else's *public key.* The private key is for decrypting, the public key for encrypting; and a computer using certain mathematical procedures ensures that neither is guessable from the other.

To see what a difference that makes, imagine that encryption procedures are locks. To send a message in a secret-key system, both Alice and Bob must have copies of the key opening the lock. Alice writes a message, puts it in a box, locks the box, then sends the locked box to Bob. Bob then uses his duplicate of the key to open the box and read the message. But in a public-key system, Alice and Bob each have their own separate private keys, and they each have their own public locks that those keys open. These locks are like padlocks: anyone can snap them shut, but only their owner can open them again. So, for example, although anyone at all can send anyone else a box locked with Bob's lock, only Bob can open it.

Imagine now that all the users of a public-key system are in a large room with a wall full of padlocks attached to a bulletin board. Each padlock has a label identifying its owner, and there are many copies of each padlock. Anyone can walk up to the wall, pull down any padlock, and lock a box containing some secrets with it. But only the padlock's owner can reopen that box. So to send a message to Bob, Alice writes the message, puts it in a box, locks the box with a copy of Bob's public lock, then sends the locked box to Bob. Bob then uses his private key to unlock the box and read the message. It's as if we all have our own personal mailbox, which only we can open but which anyone can drop mail into.

Such a seemingly innocuous system has consequences. First, Bob and Alice no longer need either a secure channel or a shared secret key. Everyone in the room could know that Bob and Alice are exchanging secret messages. Second, we now need only as many locks as there are people, rather than a lock for each pair of people. A thousand of us would need only a thousand locks, rather than half a million. Third, even Alice can't decrypt her own encrypted message to Bob; once she locks a box with his lock, not even she can open it. Fourth, Bob and Alice don't even have to know or trust each other.

Of course, computers are essential here—to make it impossible for anyone to derive information about a user's private key given the user's public lock. That wouldn't be true of real padlocks, because we could take them apart and so create a key to open them. But as far as we know now, computers make it virtually impossible to do that. We think.

The Real McCoy

Besides keeping our secrets secret, we often must make some of them public. We have to do so whenever we need to prove something to someone else.

For example, we usually tell each other apart by personality characteristics: faces at meetings, voices over the telephone, signatures on checks. Consciously or unconsciously, we test each other's identity through something that only one specific person could possess: a face, a mannerism, a laugh, a voice, a walk, a signature, a memory, a fact.

But except for the last, tests like these are irrelevant when our communications are electronic, because they can all be forged electronically. Computers make impersonation easier, particularly when our only communications are electronic. Today, as more and more information is becoming electronic, how can we prove that we are who we say we are electronically?

The obvious answer is to issue everyone identification numbers or passwords. But although the practice is widespread today, it's a very bad idea. Most of us can't easily remember a random string of letters or digits; so either we choose one that's easy to remember (a birthday, say), or we carry the number with the thing it controls (credit card, bank card, phone card, door card—whatever). The problem is that if something is easy to remember, it's usually also easy to guess. If something is hard to remember, usually we write it down and carry it around, risking its loss.

All of which brings us to a sad truth: We are the weakest link in every security system. We're both predictable and careless. We put door keys under doormats, carry our identification numbers with our credit cards, write down our passwords near our computers, and use easy-to-guess keys. We think ourselves clever when we use computer passwords like *genius, password,* and *logon,* or our names, friends' names, other common names, birthdays, license plate numbers, and other obvious keys. Ideally, we should use passwords or authorization numbers like toothbrushes: never lend them out and change them every few months.

But even if we really do memorize an unguessable number, we still aren't safe. Unscrupulous employees in the number-issuing organization could use their knowledge of the number to raid our assets. American, Canadian, Mexican, and British cash cards for example, only have a four- to six-digit "secret" identifier. That's easy to break by anyone who works in the bank's computer center. Even fraud artists who don't work for a bank can raid accounts by "dumpster diving"—going through trash bins looking for discarded statements and receipts—or "shoulder surfing"—using a hidden camera or loitering near cash machines to learn identifiers.

Cheats who only want to make free telephone calls (perhaps to complete untraceable drug deals) and who can't be bothered mucking around in our garbage, can buy an electronic scanner and listen for nearby cellular phone calls. Every cellular phone in a certain area has a "secret" four-digit identifier the company uses to determine whether a call is valid; electronic scanners can pick out those digits when a call is initiated. Once they get this identifier, impostors can make calls billable to us by modifying their own cellular phones.

Finally, anyone who can use a computer competently can commit fraud from home or office just by obtaining certain common identifiers. For example, many American companies routinely (and illegally) ask for social security numbers as proofs of identity. And anyone who knows your social security number can control your life.

Moreover, it's easy for the computer-literate among us to generate credit card numbers, because they are constructed by using a fixed mathematical procedure. Having generated, say, a hundred numbers, I can call a credit bureau and ask to verify a number. Having found a number in actual service—let's say yours—I could then call your bank and, giving your social security number (or address details, or mother's maiden name, or whatever else the bank uses to check identity), ask for a billing address change. Using the changed address, I could then charge purchases against your credit card, issue myself new credit cards or checkbooks and withdraw cash from your account.

American Express estimates that by 1993 worldwide credit card fraud alone exceeded a thousand million dollars a year, and it's increasing at 20 percent a year. Further, no defrauded company is keen to let the public know it's been defrauded. Our loss of confidence in the company would cost it far more than the stolen income. (Would you keep your money in a bank that you know has lost millions every year?) So companies often keep the figures to themselves, passing on the losses to the valid users of the system in the form of higher rates. Reported losses might be merely the tip of the fraud iceberg.

But fraud is only part of the problem. Issuing everyone a unique identifier would mean the end of personal privacy and, perhaps, the beginning of the thought police. For example, there's a growing movement today to replace paper cash with electronic cash because, with today's cheap full-color copiers, it's becoming far too easy to electronically scan and copy paper money. But cash has an advantage that credit cards don't—it is effectively untraceable.

If your bank (or the government) keeps tabs on the things you buy and when you buy them, it can learn a lot about your habits, and perhaps sell or otherwise exploit that information. To buy something, it should be enough for you to prove to your bank that you're a valid customer and that you have enough money in the bank to cover your purchases. Banks don't really need to know which of its customers buy which things.

We appear to be completely boxed in. There seems to be no way to prove that we are who we say we are without giving away enough information to let someone impersonate us later. Or is there?

Out of the Box

Renaissance mathematicians lived in the same box. They wanted to prove that they could prove something, but they didn't want rivals to know how and so claim the credit. One scheme they tried was to deposit their new proof with impartial third parties. Unfortunately, those third parties weren't always so impartial.

Nor is that some ancient dilemma restricted to mathematicians and scientists. The solution is worth serious money. For instance, when one company wants to buy another, it needs someone to handle the financial and legal details of the takeover; so it hires an investment firm specializing in that sort of thing. That firm is supposed to keep its information secret, because if news of the proposed takeover leaks out, the price of the target company's stock will soar. Insiders trading on their knowledge that the takeover will definitely occur can make an awful lot of money. Which is sometimes what happens.

What we really want is a way to divulge one piece of information without having to divulge everything connected with it. Happily, we now know, at least in theory, how to prove a fact while giving away absolutely nothing else besides our knowledge of it.

Suppose, for instance, I am a Renaissance mathematician desperate to prove to a duke that I could solve some problem, say all cubic equations. However, I can't just tell him my method because then he, or his court mathematician, may steal it. Here's what I do: I tell the duke to choose any cubic equation he wants to. Once he does, I quickly give him the solution. He can easily check my answer by plugging it into the equation he originally chose.

Of course, he's still suspicious. Maybe I just made a lucky guess. So I encourage him to pick another equation—which I then briskly solve. I keep doing that until he believes that I can indeed solve any of them. He will eventually give up and believe me because I'm very unlikely to keep guessing the right answer without having the knowledge I claim to have. Best of all, even after he believes me, he still hasn't the smallest idea of how I came by my knowledge. I've convinced him that I know a secret without actually having to tell him the secret.

To make that scheme work, four things must be true. First, the duke must have a great many possible equation choices (so that I couldn't precompute every answer). Second, his question choices must be random (so I couldn't predict his next question and solve it beforehand). Third, he must have an easy way to check each answer (to avoid giving me enough time between answers to compute much). Fourth, there must be no way to deduce anything about my secret method from any number of my public answers.

If all four conditions are met, it doesn't matter if anyone overhears the proof of identity. Watching all those questions and answers whiz by is a complete and utter waste of time since each new proof of identity will consist of new answers to a new set of randomly chosen questions. The duke can't cheat me any more than I can cheat the duke.

The important thing about this sort of scheme is that I can use it as proof of my identity without having to use fixed, and forgeable, identifiers. Suppose, for example, that I want to travel to a foreign city and be kept in the lap of luxury to which I have become accustomed. I tell my duke to write to the duke in the other city telling him that I'm coming and that to prove I am who I say I am I will instantly solve any cubic equation he chooses. I'm completely safe because no one else can impersonate me—not even my own pet duke.

It's as if two spies were meeting for the first time and exchanging the usual codes and countercodes to establish their identities. The first might say "Are the azaleas blooming today?" and the second might reply "Oh yes, but the midnight mail is quite strange in these parts, isn't it?" Or some such nonsense. However, in the updated version, the second spy has no clue what the first spy will ask—beyond knowing the general topic; and the first spy has no idea what answer the second spy will give—until the first spy gives it, at which time it can be checked. Further, the seemingly senseless question-and-answer cycle can go on for millions of times, each time with a brand new question.

The spies' abilities to survive that cycle are their individual signatures, proof that they are who they say they are. Even after those signatures are used many times—and no matter how many people listen in—no one, not even the person verifying the spies' identity, can later impersonate either of them.

Of course, none of us can go through that protocol fast enough. We can, however, put the whole thing on a *smartcard*, a credit-card-sized computer, and have the smartcard do the work for us. The smartcard could, for instance, talk to our cellular phone company to initiate a secure call—or it could talk to our computer to initiate a secure computer session. Used this way, it would become an *electronic signature* and could carry all sorts of other useful information—for example, complete medical histories, eventually including even X-rays, in case of accident.

Similar protocols can protect all sorts of other information, so that we could, for example, make anonymous smartcard purchases. No one—including the bank—has to know exactly what we buy, even if we don't use cash. We could protect our medical records and other sensitive information in the same way. The computer needn't be the Big Brother that so many fear it to be. Of course, what it could be, and what it will be, are entirely different things.

Protecting Us from Ourselves

In 1973, one particularly secret branch of the American government, the National Security Agency, was asked to help establish a standard encryption scheme for nationwide use. Along with the supersecret National Reconnaissance Organization, which mostly handles America's spy satellites, the shadowy National Security Agency is one of the most clandestine parts of the American government. Charged with protecting America's secrets and penetrating foreign secrets, it was created surreptitiously, without congressional debate, by President Truman on Monday, December 29, 1952. For many years, the government even denied that it existed. Over forty years later, even the presidential memorandum authorizing its creation remains top secret. Not until Sunday, March 9, 1991, did the agency even deign to put up a sign in front of its headquarters in Fort Meade, Maryland.

The agency—which measures the number of its computers in acres and has, essentially, a city of fifty thousand people all to itself—reputedly has more computers than any other group on the planet. It allegedly produces twenty tons of classified waste per day and employs more high-powered mathematicians than anyone else in the world. It's much more secretive than, say, the Central Intelligence Agency, and has, it is rumored, ten times the budget. It's so secret that even the number of its employees is hidden. Some say that its initials NSA mean "Never say anything"; others say they stand for "No such agency."

Of all the contenders for a standard encryption scheme back in 1973, the agency eventually chose one designed at IBM, which had invested seventeen person-years trying, unsuccessfully, to break it.

For fifteen years or more, some critics contended that IBM had been forced to put in the equivalent of a skeleton key to let the agency decrypt any encrypted information in the system without having a secret key. This belief was based on the secrecy surrounding the analysis of certain important parts of the system. Over two decades later, that secrecy is still intact. Critics also believed that the key length was too short. Although

IBM had originally suggested a much longer one, the agency vetoed it. Of course, that made many critics even more suspicious. Many believed the agency wanted keys long enough for normal security but short enough for the government to break.

Yet despite those protests, nobody publicly broke the system, and in 1977 it became the United States Data Encryption Standard. Since then, it's received massive government funding and, despite many attacks, has apparently remained unbroken. Many institutions—banks, insurance companies, stock exchanges, hospitals—use it daily, rumors of insufficient key length and possible skeleton keys notwithstanding.

Fortunately, we now know that longer keys wouldn't make it significantly more secure, so it's possible that the government didn't seriously weaken it, if weaken it they did. Nonetheless, no one outside the highest government circles knows whether the system has a skeleton key. Curiously enough, the U.S. Department of Defense has never adopted it.

Treasons, Stratagems, and Spoils

Governments are always watching, and they get jittery whenever their populations make it hard to know what's going on. For example, in 1977, after creating the first workable public-key system, computer professionals Adi Shamir, Ronald Rivest, and Leonard Adleman, tried to present it at an international conference. But a National Security Agency employee—the agency claims he acted alone—warned them that doing so would be in possible violation of the 1954 Munitions Control Act. Three years later Adleman was denied research funding because the agency feared the "national security implications" of his work.

Both cases were later settled by Admiral Bobby Inman, then head of the agency, who asked that American civilian secrecy researchers voluntarily submit their papers for vetting before publication. Inman warned that failure to cooperate would force the agency to resort to legal action.

Then, in 1986, Shamir and others devised another breakthrough, workable electronic signatures. When they applied for a United States patent, however, it was denied. Further, at the request of the U.S. Army, the Patent Office issued a secrecy order, informing Shamir that disclosure of the subject matter would be detrimental to national security.

By then though it was already too late. During the summer of 1986 Shamir and his co-authors, typical academics, had blabbed about their amazing piece of research at several universities and conferences in America, Europe, and Israel. After all, the whole point of being an academic is to think and talk (sometimes even in that order). After he received the secrecy order, Shamir wrote to his colleagues informing them of their legal obligations: "[Destroy] all copies of the paper . . . [warn] all people involved about the secrecy order." The penalty for disclosing the "secret" was a ten-thousand-dollar fine, up to two years in jail, or both.

A defense was quickly prepared. Several well-placed telephone calls were made and reporters were briefed. Everyone knew that the Army had to back down; the work had been done in Israel, with Israeli funding, by three Israeli nationals. The story was set to hit the front page of the *New York Times* when, behind the scenes, the National Security Agency again stepped in. Within two days, the secrecy order, supposedly issued at the Army's request, was rescinded. Officially, the agency made no comment.

That story is but one of many. One researcher, for example, was threatened with a felony charge because he copied and distributed declassified National Security Agency documents he had found in a university library. Now that anyone knowledgeable about computers can have absolute privacy if they choose it, governments around the world seem eager to give the impression that it's a crime to even talk about secrecy systems. And that may not be such a good thing for the future of free societies.

Too Many Secrets

If I can impersonate you, I can destroy your life. With the right identifiers, I can sign contracts in your name, empty your bank accounts, turn you into a bad credit risk, alter your health insurance, cancel your life insurance, and commit crimes in your name. To the systems that govern our lives, I become you.

It used to be that locks and safes were enough to protect secrets and that the only ones seriously worried about them were spies, generals, and presidents. Not anymore. With the computer's increasing use, secrecy systems have come to stay. Nowadays all communication is rapidly going electronic. Soon, virtually every communication act except face-to-face meetings will be electronic, and even they might be electronically bugged. The computer has led us to this impasse, but the computer isn't the problem. The problem is secrecy.

We are surrounded by too many secrets—secrets stored in and protecting information in hospitals, banks, insurance companies, communications networks, psychiatric wards, and police departments. They control access to restricted areas and school grades as well as to world currency transactions and nuclear weapons, stock exchanges, and voter-registration rolls. They govern the intimate workings of the government, the mint, the military, air traffic control, food distribution, and power stations. From transportation to energy, from food to currency, from weapons to prisons, breaking into secret systems can lead to smuggling; impersonation; industrial espionage; money laundering; sabotage; terrorism; and economic, military, and political power. Now that computers control weapon systems, secrecy systems have become as important as nuclear weapons. Which is why all governments guard knowledge of them so zealously.

We all swim in an invisible sea of secrets. Visit a bank and you create a file; go to the doctor and you create a file; log on to a computer and you create a file. Transact any kind of business—or file an income tax return, or get counted in a census, or apply for a postal change of address—and you create many files. These files have long been with us, but with the computer's incursion many of the defenses that once protected them from prying eyes—whether the government's or our neighbors'—are gone, or fading fast. Making information electronic while leaving only the old-fashioned protections in place (doors, locks, safes, guns) simply makes it easier to get.

As the earlier discussion of encryption suggests, technology can protect us, but we have to know about it before we can clamor for it. Most of us today know nothing about it so we aren't requesting it from our leaders. In fact, if anything, the reverse is happening. Big business and governments know the value of information. They've been agitating for years to get at all those lovely facts. And so far they're winning.

Clear and Present Danger

In 1948, when George Orwell published his novel *1984,* there were no computers to speak of—and he wasn't much of a technologist to begin with. But now Big Brother can hire all the technologists he wants. Despite recent setbacks in a few advanced countries, all police and spy agencies keep pushing for various anti-privacy laws, arguing that knowing everything about everybody is the only way for them to fight crime, terrorism, and subversion.

Well, they're right. But some fear that with such a system in place, nothing can protect us from unscrupulous government agents. Or perhaps even a police state. On the other hand, privacy as we in advanced countries understand it today is a modern invention. A hundred years ago nobody worried about privacy—protection against highwaymen was more important. Having the sheriff know that bandits were nearby was far more important than having him know what kind of shoes we buy.

Today, however, those of us in advanced nations don't only have to worry about one potential Big Brother; there are already hundreds of thousands, perhaps millions, of Little Brothers. Now that electronic information about everyone and everything is publically available, surveillance has branched out from its ancient roots in spying and become the province of big business, organized crime, and every computer-literate person. Indeed, readily available electronic information has grown to such massive quantities that a whole new profession is arising—that of *knowledge miner.* That's why many businesses are so busily collecting information.

Many grocery stores, for example, are now trying to give their customers identification cards. Those cards seem to be like finding money in the street: use them and our grocery bills go down because we get special deals on certain items. In exchange, the store is gaining intimate knowledge about us and our shopping habits and, indirectly, everyone's shopping habits. Of course, if we use a credit or debit card the store doesn't have to give us a card of their own—we've already given them all that information for free. That information can be transmuted into money.

Nowadays, there are an estimated five million electronic information bases worldwide, and the total volume of electronic information is doubling roughly every two years. It has already surpassed the amount of information stored on paper. By making such huge amounts of information available cheaply and quickly, and by letting anyone analyze that information, computers let us combine lots of previously useless information in new ways.

If you give me your phone number today, for example, you may also be giving me your address. In the old days I could only derive one from the other if I had lots of money. I would have had to pay an army to enter all the telephone numbers and addresses, buy high-priced computers, and hire even pricier programmers to organize and access that mass of information. Nobody but the ultrarich could do that; and they didn't need to do it because they could just hire detectives to follow you around.

But today, for just eighty dollars, I can buy two computer disks listing ninety million American homes and businesses. As soon as I know your telephone number, I also know your address. And the technology is getting cheaper by the minute.

If I know what I'm doing, here's how it might work today when you call me. Immediately I might see a map of where you live, with the location narrowed down to something like a block, or even a floor of a building. Because census information is already electronically available, I might get a feel for your neighborhood, your income level, perhaps your credit history, and so on. If motor vehicle information bases are tied in, I might learn your license plate number and registration, insurance premiums, and rate of renewal. Mail order companies and retail stores pool and sell their information; so I might also know about your annual expenditures, catalogs you receive, how much you spend on groceries, and all sorts of other things. You can learn a lot just by watching what people buy and where they live.

Perhaps you don't mind one person knowing your credit history; another knowing where you live; a third knowing when you're away on vacation; or a fourth knowing whether you're female, live alone, are elderly or ill, have ever taken self-defense classes, own a gun or dog, or have a home protection system. But if anyone can easily learn all those things just from your license number, credit card number, social security number—or even your phone number—you're simply fruit ripe for the picking.

Turning and Turning in the Widening Gyre

Liberty cannot be preserved without a general knowledge among the people. . . .
The preservation of the means of knowledge among the lowest ranks is of more
importance to the public than all the property of all the rich men.
John Adams, *Dissertation on the Canon and Federal Law*

We have erected safeguards against anyone knowing how we vote because we've had thousands of years to experience what happens when we don't protect that information. We know that voting information can be easily misused, so we don't even let our own governments have it. Until now, however, we've had no such experience with letting anyone know what we buy, what we read, what we eat, what we do, how much we earn, where we live, what we think. Most of us, ignorant of privacy systems now available, give all that information away.

Information is money. If you know a little about many people, or a lot about some people, you can make more money than if you don't. Information is a commodity. It's something you can buy and sell, mine and refine, repackage and resell, just like any material resource. Now that the secret design of a company's next-generation product is worth far more than the raw materials needed to build it, information will become the oil and wheat and iron of the twenty-first century.

Information is also power—the strongest drug in the world. All governments run on wheels within wheels that they rarely let their own populace see. To protect reputations, to win elections, or simply through greed, career-minded officials have been known to bury damaging information. To gain a political edge, even friendly nations spy on each other; and to gain an economic edge, unfriendly nations arm each other. Power has always been a game of politics over justice, propaganda over truth, expediency over morality. Nations are no more moral than the people they represent.

Because many of us dislike hearing such things, all governments find it easier simply not to tell us. That pantomime practicality, that difference between what-we-say and what-we-do, creates and perpetuates the need for secrecy. It won't ever go away.

But while the game may not go away, technology can change its rules. Today's encryption technology could, if used widely enough, make us the last generation ever to have to fear for our privacy. On the other hand, if misused, it could make us the last generation with any notion of personal privacy at all. Our choices, it seems, are growing starker and starker.

We have come a long technological way from the ancient Egyptians, yet their lives were as filled with intrigue and secrets as ours are. Perhaps nothing will change the politics of information, whether it's the politics of Cleopatra and Julius Caesar, or of your cable company, bank, insurance company, police department, video store, grocery store, or government.

Our legal, governmental, and social systems are still designed for a world without computers. Our sluggish social systems, intended for the languid bygone era of only a decade ago—an era filled with filing cabinets, paper documents, and five-day mail—haven't changed to keep up. And computer technology keeps changing so fast now that perhaps they never will.

2

Infinite in All Directions

Once Chuang Tzu dreamt that he was a butterfly. He did not know that he had ever been anything but a butterfly and was content to hover from flower to flower. Suddenly he woke and found to his astonishment that he was Chuang Tzu. But it was hard to be sure whether he really was Tzu and had only dreamt that he was a butterfly or was really a butterfly and was only dreaming that he was Chuang Tzu.

Chuang Tzu

Art is not a mirror with which to reflect the world, but a hammer with which to shape it.

Vladimir Mayakovsky

Today's computer interfaces, the imaginary places where we meet and exchange information with our machines, are simply laughable compared to what's coming. Most of us are still crouched before a screen, attended by a keyboard and mouse. Occasionally we execute little curlicues with the mouse, but mostly we type—slowly—on the keyboard, feeding a trickle of information to the computer, while it spews back a niagara of information onto the screen. We don't converse with our machines, we send them telegrams.

Turning away from the computer, we then talk to a friend, using a much more sophisticated device: our bodies. Besides the actual words we hear, we unconsciously pay close and continual attention to each other's eyes, mouths, eyebrows, postures, gestures, scents, tones, pauses—on and on and on. We speak volumes simply by shifting our gaze, blinking more, or dilating our pupils. All those grunts and sighs, smiles and frowns, let us communicate superbly with each other.

We're a long way from that level of interaction with today's comput-ers. They pay almost no attention to the enormous quantities of informa-tion we continually broadcast. The software bridging the abyss between what computers are and what we would like them to be is still very shaky, and terribly incomplete. But it's getting better.

As the decades pass, computers will grow ever more competent at ex-tracting all sorts of information from our behavior. They will eventually notice when we nod and when we yawn, when we smile and when we frown—which will let us better communicate our wishes to them and improve how gracefully we use them.

Once we start using a tool extensively, it also starts using us. We long ago passed that point with cars and telephones, and now we've passed that point with computers. That fact will have consequences that go far beyond simply easing the simple jobs we use computers for today. Ultimately it may change how we view reality itself.

The Medium Is Massaged

Few of us have yet had the luxury of seeing the latest fantastic graphics on the screen of an advanced computer. The hardware is still expen-sive and the software is still crude. Even so, most of today's newspa-pers, magazines, and books have already passed through a computer at some point. Thus the most immediate effect computers have had on the graphic world has been through the printed page.

When we take a pen to paper, whatever we write or draw is the orig-inal, and a reproduction is just that: a copy. But when working on the computer there isn't any original. Every copy, whether it's the first or the millionth, is just as clear, accurate, and valid as any other.

In addition, when created on a computer, the appearance of a page of text, a drawing, a painting, or a movie is malleable. Revision in the traditional plastic arts—painting, sculpting, film—used to be hard, but no longer. On the computer, the graphic arts come to resemble a jazz performance, where the original melody is merely a set of loose instructions and the interpretation can be unique every time.

Once we have something in a computer, we gain unprecedented con-trol over it. For example, we can change, distort, or rearrange a photograph without damaging the original. With the right software, we can give Aunt Edna a haircut and see how she looks with it before making a print. This control has interesting consequences. Perhaps you just received a fashion catalog showing models wearing various articles of clothing. Everything looks real. However, some of the models may never have worn some of those clothes. The models are real, the clothes are real, but the application of the clothes to the models' bodies may have happened only inside a computer. In a few cases, the clothes may not even be real. Eventually, perhaps, the models won't be either.

Already today, after a photo shoot, fashion photographers often retire to their computers and rearrange their pictures. Faces can become wrinkle free, hair more lustrous, irises more brilliant, eye whites whiter. Eyes may move, ears may shrink, mouths may widen, and necks and legs may lengthen. Models might find that they've miraculously lost weight in various places—and gained it in others. The changes are usually subtle, but the overall effect can be to create almost a new person. Electronic surgery is far cheaper, quicker, and less painful than the real thing.

Touching up photographs is normal. But this level of manipulation is new; it's the creation of entirely new pictures. Further, to do it you don't need several thousand dollars' worth of photographic equipment or decades of photographic experience. All you need is a computer with the right software—which is getting cheaper by the minute.

Once upon a time, photojournalism was seen as the realm of objective observers, supposedly documenting the world as it was. "The camera never lies," it was said. Taking that as fact, courts still accept photographs and videotapes as evidence—although they may not for too much longer. Soon, millions of us will be able to alter any picture, seamlessly. And that has all kinds of consequences.

Refractions Through a Computer Screen

Producing realistic but fake video today takes lots of money. However, the technology is advancing briskly, the price is dropping rapidly, and the potential rewards are increasing enormously. By decade's end, fake video will be indistinguishable from the real thing. It's only a question of money.

The film industry, for example, driven by the demand for ever more spectacular special effects, is moving rapidly toward all-electronic movie production. In the 1990s, technology gave us special effects films like *Terminator 2: Judgment Day, Jurassic Park, Forrest Gump,* and *Toy Story*, all of which stunned audiences worldwide. But they are only the beginning of what will soon be possible for much less money. Within a few decades we may have the first computer-generated live-action movie. It will look as if it were shot with real cameras and real actors at real locations. Yet it will all be make-believe. Such a film may look even more real in some ways than today's camera-created movies, because the computer could simulate such things as the tiny starbursts of eyelash diffraction we see when looking into a bright light on a dark night, the rainbow-tinted view through rain-wet eyelashes on a bright day, or the subtle blurring of tears.

Of course, the film industry probably won't go in for such subtleties at first. We're more likely to first see bright, vivid colors, speedlines in car chases, and extra bright muzzle flashes for the gunfight in the last scene. But tomorrow's Fellinis, Truffauts, Kubricks, and Kurosawas are probably playing video games today. When they grow up their genius will give us art that uses the computer with the same nuance and sensitivity that today's best directors bring to film.

Such computer technology might give directors yet more power over actors. Today, Humphrey Bogart, though long-dead, is again starring in new films as computers let us extract and piece together his images from old films. But that's only a crude step along the way. With enough money and time we'll be able to create "body-maps" of dead actors that detail their speech, walk, gestures, and so on, letting us make them do scenes they never acted in.

Eventually, living actors may license their body-maps so that their voices, expressions, gestures, and body movements can be recreated in any sequence. Perhaps in the future, we might ourselves use the body-maps of famous actors just as we now drive someone else's car. A few years later, filmmakers may not even need to model their creations after real actors; by then computers might be able to make human figures with their own special characteristics out of whole cloth. Directors would then need only a few famous actors, recreating the rest of the cast from body-maps and, therefore, totally controlling them. If today's actors think the job market is tough, they have no idea what the competition might become twenty years from now.

On the other hand, as the price of the technology continues to plummet, directors will inevitably find themselves competing with actors, who could by then afford what would today be a multimillion-dollar movie studio.

What works for actors could also work for other performance artists. Instead of going to a studio to learn ballet, students may one day buy an adaptive body-mapped recording of a master going through the routine, with all movements recorded exactly. Students "driving" such a recording might then practice by putting their hands, feet, and bodies in the precise attitudes struck by the recorded master. They won't be watching a recording; they'll be inside it.

Suppose that someone copies and sells that recording. Would that be invasion of privacy? Copyright infringement? Piracy? When body images and movements become copyable information, what becomes of personhood?

News At Eleven

When computers make it cheap enough for us to generate any seemingly live-action scene we wish, we may have news programs consisting solely of simulated scenes. We already have dramatic reenactments of reported events on television, and massive amounts of computer graphics and image massaging in advertising. Considering the artificiality of television commercials today, we're already well on the way to artificial news. It's inevitable that someone somewhere will use that technology for propaganda.

Today's television sets have knobs to control their displays in primitive ways: color, brightness, contrast, and so on. But if television begins to show artificial news, perhaps we could also have a reality knob to change the realism of the scenes we view. If so, then how about a drugs, sex, and violence knob? Or a political-bias knob?

What works for television sets can also work for telephones. Today, if you want to change your telephone voice because you don't want strangers to know you're female, you can buy a device to do so. To-morrow, you may want to change your voice to make yourself sound ill when pleading sickness to your boss. Eventually, voicephones will turn into videophones. Then we'll start shopping around for cosmetic soft-ware to manipulate our images second-by-second. Being conveniently ill will always have advantages.

Eventually, we may not believe anything we see or hear on a screen, a phone, or a page. Then, perhaps, the government will mandate that the news must be real. Or perhaps not. Maybe we'll prefer our news simulated, because it looks and sounds so much better than the real thing.

All of this isn't new, of course. We've been doctoring movies for as long as we've had movies to doctor. For example, many First and Second World War newsreels were either staged or faked outright. And the fak-ing goes back at least as far as the turn of the century; many newsreels of the Boer War were shot in Jersey instead of South Africa, where the war actually took place. Even then filmmakers knew that faked action looks much more real than the real thing. In a real war, too much happens too fast for even military viewers to make much sense of it. Further, soldiers make bad actors. They have an annoying habit of fighting when it's too dark to shoot good pictures; they persist in not clumping into nice, neat bunches during attacks; and they either die too messily for stay-at-home viewers or they don't die dramatically enough.

Of course, all that changed when the Vietnam War came along, and a camera shrank enough for one person to carry and use one. Everyone could finally see exactly what war looked like, right there in the living room. Then came the 1991 Gulf War. And as the phosphorescent tracer rounds lit up the night-scoped Baghdad sky, real war became even more pyrotechnic than film war. At least for a while.

"Seeing is believing," we used to say; but that's never really been true. During the Second World War, newsreels showed Hitler dancing a jig after the fall of France in 1940. Hitler was many things, but a jig fancier he wasn't. He never did dance that jig—he just lifted his leg. It was Allied newsmakers who optically looped that leg movement into a jig. Perhaps one day governments will use computer simulations in a similar way to keep deranged or dead political leaders in power for twenty years, looking younger every day. Then we'll all be jigging to the beat.

On the Screen of a Machine

In a darkened room, a young woman sits crosslegged before a keyboard and a large glowing screen. We can see a small rectangle in the screen's upper left corner—a well-known news announcer solemnly mouths the day's news, unheard. Above the announcer blinks the date: December 20, 2006. A small Japanese kabuki mask grins in the screen's opposite corner.

The woman isn't looking at the news broadcast or at the mask. She is looking into another part of the screen, a soaring view from above of a huge green-and-yellow checkerboard of waving grain. The scene is so perfectly rendered that when she presses a key, we feel as if we are rising from the surface in a silent helicopter. The motion is smooth and perfect, with none of the jerkiness of normal motion. When it stops, we notice a tiny animal picking its way across the fields far below.

Now the woman gestures with her right hand and we see that she is moving a small curved box on the floor—a computer mouse. As her hand moves we seem to swoop down toward the animal and land some distance in front of it. It is a dinosaur. The viewpoint changes and now we are looking up at the huge surreal dinosaur—a Tyrannosaurus Rex—as it soundlessly strides by overhead casting its shadow all around us. We now see that it isn't real; it is smooth and perfect, like a giant soft plastic toy. It flexes and shifts as it glides by, its legs working with Victorian clockwork precision.

In the top left window, the famous newscaster is replaced by a cartoon, and the audio cranks itself up. "This just in," the cartoon says in a synthesized voice, grinning insanely, "Your buyer has found a pair of vases at auction in Sydney. They are available at the price you specified. Do you wish to verify your bid?"

The woman gestures with her right hand, and the dinosaur scene vanishes, revealing many smaller rectangles filled either with text or otherworldly scenes. Each window is a view onto another world. The woman types something with both hands. Suddenly the kabuki mask frowns slightly, mirroring the computer's surge of effort. A new window blinks onto the screen, showing an image of two vases sitting in a room on the other side of the world. The camera rotates slowly around the vases, showing them from all directions. The video image is blocky and slightly stuttering, clearly not as well rendered as the artificial dinosaur scene.

The woman grins and types some more. A softly lit room with angular furniture appears, displacing all the other windows on the screen. Unlike the glowing dreamlike dinosaur, this image looks like a photograph. The wood floor has the texture and color of some dark oak; it looks slightly scuffed, as if from too many people with hard shoes. Familiar paintings adorn the walls, and copper and brass accents glint on their frames.

We can see light refracting through a heavy glass table squatting in the middle of the room, breaking the lines of the chair behind it. A small brass halogen lamp casts a bright light pool on the woven rug, and the shadows cast by the table legs fade off into the room's dark corners. Then the scene changes slightly. The pool of light becomes more diffuse, and the shadowed corners of the room brighten slightly. We notice highlights and reflections on the glass. As the viewpoint changes, we gradually stop seeing the image as a photograph and start, unconsciously, thinking of it as a doorway into a real place. That particular real place, however, exists only inside a computer.

The woman appears uncertain. She reaches for a pair of thin gloves with wires trailing from them that lies next to the screen in the otherwise empty room. She waves her gloved right hand and it appears as though she has somehow reached in and placed one of the vases on the table, for we can now see sketches of her hands in the scene, moving in time with her gestures. She reaches again for something we can't see and the vase is filled with bright flowers, azaleas and chrysanthemums.

Smiling, the woman dons what looks like a pair of wraparound sunglasses—which also trails wires—and stands up. Turning around she appears not to see us; it's as if she is now in the room on the screen, for we see her cartoon image suddenly appear in it. She walks about the imaginary room, surveying her handiwork. Apparently dissatisfied, she points to one of the heavy, craftsman-style benches and it follows her gesture, scuttling over to another wall. She points again and two chairs walk by the table and snap into position around it. She snaps her fingers, says "Color cube," and a three-dimensional multicolored cube about the size of her head appears in front of her, slowly tumbling end over end in space. She points to one part of the cube and then to the flowers in the vase and they change to more subdued white roses and pale pink lilacs.

Finally satisfied with her job, the interior designer flips an imaginary switch in the air with her gloved left hand. The kabuki mask smiles, and, as it does, music wells up and the woman begins to dance in the bare room. And her image is also dancing in the richly furnished room, on the screen of the machine.

Jacking In

When we go to the movies, many of us live a vicarious life for about two hours. Our attention can sometimes be so riveted on the screen that we neglect to attend to our own direct experience. We may forget the chair we're sitting in and the people who surround us. We become immersed.

That feeling of immersion can happen with a good book, or even with a good television show, but it becomes even stronger at the movies because then we're in a darkened, unremarkable room facing a large, bright screen that demands our sole attention. Immersion becomes easier still if the action wraps around our eyes, as it does at some theme parks. The action then covers our peripheral vision. When that happens, we often forget where we are and start thinking we're somewhere else.

We can further heighten the visual illusion by using a visor with two separate tiny screens that give each eye a slightly different view of the scene. Our brain seamlessly blends the two views and gives us depth perception, so that we feel as if we're seeing a real, three-dimensional scene. The same illusion works when sounds seem to surround us; listening to a good stereo system can do it. Evolved to interpret only one reality at a time, our brain often starts thinking it's in the environment producing the sounds.

Yet another way to induce the illusion of being elsewhere is to sit in a reactive chair that moves and shakes as it would if we were in another environment—say, a helicopter flight high over the Andes, an earthquake in Los Angeles, or a rock concert at Wembley Stadium.

The whole point is to enter an immersive world. If a simulation is sophisticated enough in sight, sound, and sensation, our brain suspends disbelief in the artificiality of the experience. When that happens, we enter a new world.

Then, to complete the illusion and to move beyond simple entertainment, the immersive world could become reactive—that is, it could respond to our wishes. If a computer could notice a turn of our head, a flick of our wrist, or a blink of our eye, it could change what we sense. If that happened then what we do would affect the world. We would enter a *virtual reality.*

Reactivity alone is so powerful a seducer that video game aficionados often fall into a video game world, even if the game's designers have made little attempt to make it realistic. It is significant that video games are probably the single most popular use for computers today. Once our actions have direct effects on our environment, its sight, sound, and sensation qualities almost don't matter. When what we do makes a difference in the world, then the world, no matter how unrealistic or unrealizable, is enterable through the door of our imagination.

When this happens, we can do and be anything the computer will let us do or be. If the system were good enough, we could enter a new universe with its own laws. There would also be no reason why we couldn't link up with many others scattered around the globe, each immersed in the same artificial experience. We would then be jacked into an artificial world—a virtual world—cyberspace.

Video Games with Real Blood

For decades now, the cockpits of warplanes have come more and more to look like an explosion in a video arcade. To combat the clutter, pilots of today's most advanced warplanes now have see-through displays in front of their windscreens—or, more recently, two tiny video screens built right into their helmets. When flying at night, these video pilots no longer see reality. They see a computer-generated abstraction of the cockpit instruments, terrain, and nearby warplanes. That lets them track missile readiness, combat damage, flight information, and enemy locations while keeping an eye on what's happening outside.

For its weight and size, an advanced warplane is one of the most expensive machines on the planet. Aircraft costs have quadrupled every decade for the past eight decades, and extensively training pilots to use one is a luxury most nations simply can't afford. Over sixty years ago, therefore, advanced air forces started using (noncomputer) simulators to train their pilots. These machines tried to mimic the plane's behavior, giving the pilot a feel for what things might be like in the real cockpit. Of course, they were crude. But as computer technology matured and the expense of real planes skyrocketed, the simulators improved so much that it's now easy to believe you're in a real plane.

Besides training pilots, simulators are now also used to train ship and submarine crews, tank and artillery operators, and even missile-battery and ground attack troops. The American military, for instance, already has a few hundred tank, missile-battery, armored vehicle, and aircraft simulators linked in a global network—an entire army online. That network will soon simulate all aspects of war, from individual tank movements to the strategic maneuvering of whole armies. So tank crews in Germany could work with air crews in Saudi Arabia and artillery crews in California to fight a simulated war against their counterparts, who are also scattered across the globe.

Computers can simulate the landscape the pilot or tank operator would see if flying or driving over it—every river, every tank, every telephone pole—and they can manage the various weapons. Computers can immerse combatants in a shared but artificial battlefield despite their physical separation. The synthetic battlefield lets simulated munitions lobbed in Germany "hit" in California a few seconds later, just as if the gun and its target were near each other. Computers can even simulate a horde of enemies, basing their reactions on information about how they usually respond.

Advanced tanks are already one step away from reality anyway. Often their operators only see a computer-massaged version of the outside world through periscopes. Twiddling the knobs and dials and looking into the computer monitor to judge the effect, tank crews easily accept a simulated battle as a real one—except for the smell. Replaying the same battle in dozens of different sequences trains them to use the tank as if it were part of their own bodies so that responses become second nature. Such preparation is invaluable when real war comes.

Without combat-level practice, peacetime military forces turn to mush. Simulated war helps them weed out poor combat performers. And, while nowhere near as realistic, simulators are a lot cheaper than open-air military exercises. Armed forces that can afford simulated war are fielding near-veterans with extensive knowledge of the foe's own terrain.

Of course, simulated war will never equal real war—unless we're prepared to kill some of our soldiers during training. Nonetheless, American tank operators deep in the bowels of a plastic tank simulator bolted to the floor in Kentucky can battle each other just as if they were in the arid deserts of Saudi Arabia or the rolling hills of Germany. These troops aren't on the frontline, they're online.

The Eyes of a Stranger

Besides putting us in an artificial world, the new technology could put us in the real world—although not the one we're used to. For example, now that mobile robots are becoming competent enough and cheap enough to interact with the world in interesting, nontraditional ways, it is becoming thinkable to control them using *telepresence*. By linking the robot's senses to some of our own, and by damping out our own senses, we can temporarily fall into the belief that we are the robot. By linking our actions to the robot's actions we can then control it. The feeling is indescribable.

Imagine that you are in a room with a visor over your eyes and equipment attached to your hands. There is also an inert humanoid robot in the room. As you begin to move your head, the robot's head moves in lockstep and you immediately see the room from a new perspective. Since you can see nothing but what the robot's cameras see, you soon start thinking of yourself as being where the robot is, watching a strange person (that is, you) moving its head. Moving your arms moves the robot's arms, which soon start to feel like your own arms, while the corresponding gestures of the person across the room come to seem more and more like a puppet play. If the interaction is seamless enough, you eventually forget your own body and imagine yourself present in the robot—you become *telepresent*.

Now imagine the consequences. If you are a surgeon, say, there needn't be any connection in space between you and your patient. You could be in the same room, in the next room, or on another continent. If you are a miner, you needn't be anywhere near the ore. You could be telepresent in a robot mining riches deep in the ocean—or deep under the earth or on the moon. Nor would you have to remain your human size. You could will yourself into a kilometer-long robot body in orbit around the planet. You could operate on human cells using microtools. Or you might monitor aircraft in a few cubic kilometers of airport airspace.

These and other uses might become common in the decades ahead. Soon we might even have telepresence toys for children. Junior explorers could have a tiny robot equipped with tiny cameras and a radio link. They might use it to explore the garden through the eyes of a mouse. As the technology progresses, the perspective could eventually shrink to a worm's eye view, then a microbe's. These robots in the playpen might prepare children for the multiplying telepresence tools available to general society. Reality will never be the same again.

Transcontinental Tennis, Anyone?

It's eleven o'clock on the evening of December 20, 2011. The obsequious attendant, blandly handsome in that ill-defined way brought on by too much cosmetic surgery, hands you the tennis racket, wishes you a good game, and withdraws. You make a note to yourself to try this again sometime; the cost is certainly worth it for the flattery value alone. It's much better than a simple videocall. You heft the racket; it feels real. The weight is right and you can even feel tiny ridges in the racket's handle. The muscles in your arm and shoulder bunch and flow with the slight effort as you swing it, limbering up. You can hear air whistling through the strings in the moving racket.

A nod to your opponent—who smiles at your visual exploration of the court, the glade, the racket, and the ball—and the game begins. You feel the gentle breeze the attendant has told you will increase to cool you as you start to sweat. It is computer controlled. The birds in the trees surrounding the glade continue their warbling despite the gentle wop of the ball sailing back and forth. They too are computer controlled.

Only a few things are different from your usual game at the local tennis club in Chesham. Your eyes and skin tell you you're bathed in warm sunshine, yet you know it's night in England. Also, your ears aren't reporting the muted traffic noises you should be hearing. Even your nose may catch the scent of fresh cut grass and traces of windblown dust; yet you know you're indoors. The experience is real to you. Even so, your tennis racket, the ball you're now reflectively bouncing against the grass, the court, the net, the warm sunshine, the breeze, the birds, the trees, the glade—even the attendant—all are unreal.

Only your opponent is real. But she's currently visiting Vladivostok, ten time zones away, taking a break from the trade talks that morning to play a quick game with you back in England. The image you see of her, delayed less than a fifth of a second despite the great distance and enormous computations, is computer generated. Maybe she's no more real than the attendant.

Beyond the Looking Glass

Any electronic display that knows where you're looking can be a window on another world. As you move in front of the display, the scene could change to let you see things as if they were just behind the display. For example, the windows of your future house needn't display the same scene day after day. On Monday you might want to look out on the Himalayas; on Tuesday you might prefer the beach at Waikiki; and on Wednesday you might like to see your neighbors as they appeared on Monday—or as they appear right now—or a year ago last Friday. Perhaps it is night and you want to switch to an infrared view of your neighborhood to check for prowlers. Or maybe you want to zoom in on some detail outside that you can't quite make out.

Such smart windows needn't be static two-dimensional views like paintings; they can be windscreens looking onto another place or time. A glass cube in your home could be an electronic aquarium, full only of electronic fish; and when you grow tired of that, it could become a television set. And what works for wall windows can also work for portable windows. Hold up such a window and you hold a view onto another world—any world. Once the computer knows precisely where you're looking, it can change the display as your gaze shifts.

Alternatively, the screen may pay no attention to your gaze but display something new whenever you move it. You might sit in a swivel chair holding such a display in front of you and, as you spin in the chair, get a three-hundred-and-sixty-degree view of another world. Maybe today you want to see the Grand Canyon from the inside, view an abstraction of all the electronic information in your office, or look through the windscreen of a fighter plane banking over the Andes. You might choose a view of the normal world with electronic overlays that explain things—perhaps a broken car engine you need to fix, a patient's body you're about to operate on, or a book you're looking for at the library.

Shrink the display and you end up with a smart photograph. When first seen, it looks like today's static, single-view photographs; but once you move your head—or move the display—the photograph changes to show you what you would see if you were looking at the scene from that new angle. You have, essentially, a hologram of the scene. Each viewing angle shows you something new.

Or consider mirrors. Mirrors today only show you face front. Moreover, they mirror-reverse you, you must always be looking at them, and you have no way to compare yourself over the years or wearing different clothes. A smart mirror—that is, a camera, a computer, and a display screen—needn't have any of these limitations.

Some expensive cars today already carry primitive forerunners of devices like these. They have a dashboard display screen showing the road system and the path to be followed, updated second by second as the car moves. They also have cameras in their skins and dashboard displays showing views from behind the car—or directly in front of it—freeing the driver from having to use awkward rearview or sideview mirrors.

Mirrors, photographs, books, aquariums, windows, walls—perhaps even entire rooms for the very rich—all might soon become windows onto new, active, three-dimensional worlds.

The Imagination Set Free

Computers and the reality-distortion field they induce are edging closer and closer to our skin. First, they squatted in their own rooms on their own floors. Then they jumped onto our desks and briskly evolved into machines that commute with ease. Palm-sized computers have already become far more than mere toys, and there are plans in the works for wearable computers. Can washable computers be far behind?

To interact with our computers better we're shrinking the bulky headgear we have to use to see and hear alternate realities down to something the size and weight of a pair of sunglasses—very expensive sunglasses. Eventually, they might shrink to the size of contact lenses. As computer power improves, the bulky handgear we must now wear to let the computer decipher our gestures will also eventually vanish, to be replaced by tiny video cameras or laser beams that let the computer monitor all our movements—facial expressions included—simply by analyzing images.

Once we've passed that point, perhaps fifteen years in the future for cheap systems, the technology might shrink even more and eventually slip inside our bodies, just as pacemakers and birth control implants already have. As computers continue to shrink in size and cost and grow in power and complexity, we might one day use neural implants to interact directly with our machines. Two decades from now, some of us may start choosing to be artificially augmented. We would then be attached to what would truly be our very personal computers.

Electronic cochleas that let the deaf hear already exist. Of course, they are still crude. It takes at least a month of healing after implantation, then several weeks of adjustment to the new stimuli before patients can hear with them. But by using those devices, otherwise deaf patients can recover 30 to 40 percent of their hearing.

Artificial nerves also already exist. Some paraplegics have had thin wires, which are connected to a tiny computer in their abdomen, implanted in their paralyzed legs. Flexing various of their upper body muscles tells the implanted computer to apply power to various leg muscles. Within a few months of training, patients who were once told they would live in a wheelchair for the rest of their lives walk again.

It is already becoming practical to electronically read the signals carried by individual nerve fibers. Researchers have made permanent connections to the leg nerves of rats by cutting the nerve, inserting a chip with about a thousand tiny holes between the cut ends, then letting the nerve fibers regrow through the chip's holes. With such an implanted chip, they can read individual pulses traveling up and down the nerve fibers and send artificial pulses down the fibers to move muscles. Such implants could eventually give amputees lifelike control over their computerized artificial limbs. Eventually, computer implants might even give us manual remote control of all sorts of objects over long distances or at small scales.

In the far future—thirty years ahead, perhaps—we might not even need to invade our bodies to gain these powers. Even now brain scanners can detect minuscule electrical currents in our brain and use them to control machines with our thoughts. Various military units are already experimenting with controlling warcraft mockups with such brain scanners. Controllers simply think about moving the mockup in a certain way and, using biofeedback, a display helps train them to influence the mockup's behavior.

Of course, all that technology is still primitive—and very expensive. One day, however, the disabled may use it to direct their wheelchairs, or their own disabled bodies. The blind and the deaf might use it to see and hear again; and armed forces might use it to control jets and ships and tanks. Ten years later, all of us may use it to guide our vehicles and household appliances.

From such technology we might build strength-amplifier suits for soldiers, or for miners and workers in other heavy industries. Perhaps we will use it to aid our memories and enhance our intelligence. When we have artificial eyes and ears, it's but a step to giving ourselves artificial memories so that we can replay past scenes and recall exactly what someone said or did.

Further, when we can block normal nerve impulses, a computer could replace those pulses entirely and so create a totally synthetic reality. Whatever impulses we send to our muscles could be used for any other purpose. You might jack into your car and receive kinesthetic feedback on its condition. But it need not stop there. Future executives might jack into their companies just as they jack into their cars. When that kind of technology comes, it will no longer be *virtual* reality, it will be *real* reality.

Direct interaction with machinery, for example a car, could speed our reaction times by eliminating the time it takes to move our muscles. We could also use it to prevent sensory overload in pilots by giving them more direct control over their planes and removing clutter from the cockpit. Or we might use it to put a display in our field of view that we could use for everything from annotating sales records to aiming a tank gun.

Eventually we might modify our pets and ourselves to carry all sorts of technological enhancements directly hooked into our nervous systems. For example, in the future some of us might see heat or ultraviolet light as easily as we see visible light today. Imagine what that might do for the security industry alone. Imagine guards who can shoot guns they don't even hold. Imagine cyborg guard dogs armed with grenade launchers.

All these abilities will trigger yet another change in our perception of reality, perhaps the last one. This time it won't be simply a change in perception or memory or acuity or skill, but a change in reality itself. Perhaps our deepest distinction is that between our own bodies and our environment—the self and other—and that distinction crumbles when we can jack ourselves into any device in our environment. In such a world, the environment becomes us and we become the environment.

In forty to fifty years, the best, most expensive, alternate realities might become as detailed as reality itself. There might be no way to tell the difference—at least for sight, sound, temperature sensation, motion, and (perhaps) touch. Smell might take longer, not because it's impossible but because it may be expensive; and unless we go to neural implants, we may never get taste at all.

The question "Is it real?" might first become irrelevant, then meaningless. The use of synthetic experience may grow so widespread that some of us will have to make a conscious effort to act as if reality were real. For some of us, everyday reality might become just another alternative reality, one whose natives take unkindly to being treated as if they were figments of someone's—or something's—imagination. Perhaps a new fashion for unmediated realism will arise, a backlash from all the artificiality we live in. Or perhaps not.

Will we, in the very far future, use neural implants to brainwash people, record dreams, read minds, exchange experiences directly with each other, and give ourselves endless pleasure and prisoners continuous pain? A strange new world is coming, and coming fast.

It is unknown—and unknowable. We're headed for it at breakneck speed, accelerating as we go. But we don't know where we're going. About all we can be sure of is that it probably won't be any of the cozy futures we think we spy today as we fashion our own little bits of it. It will probably be something wholly other, something alien, because that's what this level of corrosive change does to things. It mutates them into something rich and strange. Lacking any formal way to predict the future we're now so busy crafting for ourselves, we can rely only on poetry and metaphor to draw its lineaments and encompass its strangeness.

Of course, all this—even if it does come to pass—isn't something we have to think about today. It's all unthinkably far in the future.

Isn't it?

In the Screen of the Machine

Every culture gets the magic it deserves.
Dudley Young, *Origins of the Sacred*

"You have a message from Fujiko, sir," Jeeves, his supervisor program, tells him in its Oxford accent.

He looks up. He is working in one of his virtual offices, but Jeeves has standing orders to interrupt him for Fujiko's messages. Jeeves's image—for the last few months dressed as an English butler—is holding a gold tray with a note on it. "Read the message, Jeeves," he says tiredly.

Jeeves's image picks up the simulated note and pretends to read it. "Home for lunch in an hour. Fujiko."

The man smiles. As usual, Fujiko didn't waste words.

Jeeves coughs gently and dematerializes the tray and note. It seems annoyed at the message's abruptness. "If I may take the liberty, sir, over the past three hours Secretary received ninety-two other messages. Thirteen were prerecorded holocalls, fifteen were text and pictures only, four were voice . . . "

"I'm sure if any were urgent you would've interrupted me, Jeeves."

"Just as you say, sir," Jeeves says in its don't-interrupt voice. "Besides the calls, News has found 604 potentially interesting . . . "

He waves Jeeves to silence. "Thank you, Jeeves, they can wait." Two-year-old Jeeves is already getting quite persnickety. When he first bought the program, it was as bland as American cheese. But as it found a persona to assume, it—or should that be he?—was rapidly becoming an aged Gruyère.

"Please send the news to Library for integration and summarize the calls and mail to one sentence descriptors. I'll listen to it later." Jeeves, managing to look both pleased and miffed, nods and obligingly vanishes.

Sighing, the man wonders just how Jeeves spends its computer time when it doesn't have to service his requests—which is most of the time. Does it have a secret life? Smiling now as he tries to imagine what that could be, he speaks an activation word and his present synthetic office vanishes, to be replaced by another.

He now appears to be lying on a couch next to a coffee table in a small transparent room, as if he is inside a giant soap bubble. Through it he sees a moonscape with a robot miner at work—one of his whimsies this month. Just above his head, a line of what appear to be paper books floats in mid air—a small window onto the world's library of electronic books. There is a clutter of small objects on his simulated coffee table: a fold-down paper calendar with a ticking clockface embedded in today's date, a tiny file cabinet, a flower in a vase, and a small toolbox. There is no keyboard, screen, or telephone.

What looks like a paper calendar is really an agenda file, and the clock is an electronic simulation that tells him about upcoming appointments. The tiny file cabinet is an enormous filing system, which contains electronic, not paper files. The health of the flower reflects the state of his business. Its color, the shape of the vase, and the amount of water in it reflect orders, contracts, and service requests. It gives him a quick way to estimate how well he's doing. Right now it is blooming in a tall thin crystal vase with plenty of water.

Ignoring the flower, he commands the toolbox to open and it does, revealing some small shimmering objects. He picks up what looks like a waterstained cardboard box, places it on the coffee table next to the toolbox, and tells it to open. He can see some tiny objects inside, and when he directs them to grow they do.

The man starts rummaging through the things at the bottom of the box. For the most part, the objects look like chrome or pewter widgets, differently shaped and colored; in fact they are representations of his neural-probe programs, neurophysiology software models, pharmacology texts, psychology toolboxes, old brain scans . . .

After a few seconds of rummaging, he looks a bit exasperated and says, "Jeeves, find me my latest brain simulator prototype." Immediately a cream-colored pillbox hidden in a corner starts blinking red; then, through a complicated origami trick, all the other things—including the coffee-table—fold into themselves and shrink away as the pillbox grows to handy proportions.

Soon the clock bleats softly and, using Jeeves's voice, tells him it's time for his next meeting. Snapping his fingers, he says "Shift me" and immediately pops up to High Hilton to meet with a Neurotek representative and go over some initial designs. His comfortable virtual office has vanished entirely, and although he knows exactly where his body is, he is still a bit disoriented. Everything he sees and hears tells him he is now a sophisticated motorized camera bolted to the wall of a cabin in orbit far above. He is now inside a reality, inside a reality, inside a reality.

When he concludes the meeting fifteen minutes later, he speaks the word to power down his electronic contact lenses. The orbiting cabin vanishes, leaving the couch on the moon, which also vanishes, leaving him back in his first virtual office, which also vanishes.

His physical home finally opens out on him. After spending the entire morning in various artificial realities, he's finally back in real reality—such as it is. He is alone at home, swaying gently in a hammock on a veranda facing out to the sea near Kōchi Bay. His computer is in his shirt pocket; its microlasers, which tracked his gestures and facial expressions and sent information to his earplugs and the tiny displays in his contact lenses, wink out.

As usual the transition has left him a little unsettled. His brain, evolved to deal with one stable reality, still isn't used to abrupt jumps between realities. Someone should fix that, he thinks. It would be even worse when computers talk directly to the brain. He would have to remember to make the transitions in his new interface more graceful. He rubs his eyes tiredly—being careful of his contacts—and thinks of the years of work ahead for him and his people.

Neurotek had been very insistent at the meeting. They want the first of the next generation of interfaces, direct-to-brain interaction, in three years. It would be crude, of course; the first generation of anything always was. Neurotek knew that the erotica market alone was worth killing for. If they got what they wanted, they would be years ahead of the competition, and reality would become even more unreal.

What is real? he thinks. Some of his older programs are already becoming more real to him than many of his childhood friends. And his alternative realities are becoming more pervasive every day. Sometimes he catches himself snapping his fingers and uttering his various call words, unthinkingly trying to summon things in his real home. They rarely come of course. It's December 20, 2021, and reality isn't by any stretch of the imagination what it used to be.

His stomach rumbles, interrupting his thoughts. Well, at least that's real. Never mind the philosophy. He would test his first mockup after lunch. Fujiko would be home soon.

3

The Power of Ideas

A book is a machine to think with.

I. A. Richards, *Principles of Literary Criticism*

A stand can be made against invasion by an army;
no stand can be made against invasion by an idea.

Victor Hugo, *The History of a Crime*

If nature has made any one thing less susceptible than all others of exclusive property, it is the action of the thinking power called an idea, which an individual may exclusively possess as long as he keeps it to himself.

Thomas Jefferson

They were kings of the hill. They dominated every ecosystem. But for them, as for all life, there is only one rule: adapt or die. As they roamed the veldt and the marshes and the seas, living life as they had for millions of years, an asteroid was just entering the atmosphere. Things were about to change greatly. And the dinosaurs had no idea.

The asteroid was about ten kilometers across—the size of a small mountain. Traveling at over a hundred thousand kilometers an hour, it would strike with an impact releasing about a hundred million megatons of energy. The destruction would be roughly equal to exploding ten copies of the bomb used to destroy Nagasaki on every square kilometer of the earth's surface, oceans included.

As the pitted mass of iron and nickel entered the atmosphere, any dinosaurs looking up would have briefly seen a giant fireball. First it flared orange, then blue-white, then became more blinding than the sun—all in fractions of a second. It may have hit just north of the Yucatán Peninsula in Mexico. If so, it simply brushed aside the ocean and slammed through to the seabed, then straight down into the magma beneath the earth's crust. There it vaporized in a titanic fireball extending far into the stratosphere. The local temperature jumped to ten million degrees centigrade and the sky began to burn.

The impact produced an instant volcano, cratering the earth eighty kilometers deep—deeper than the height of twenty Everests—in a circle two hundred kilometers wide—an area bigger than Vancouver Island. The blast heat vaporized perhaps a million million tons of seawater and made silt and salt rain down all over the world.

A shock wave as dense as a wall of granite ran through the oceans and the earth's crust at nearly the speed of sound, crushing everything in its path. Kilometer-high tidal waves, monster hurricanes, worldwide firestorms, earthquakes, and volcanoes followed in its wake. All the earth turned dark as clouds and dust pinched off the life-giving sunlight. Dirty salty rain and snow fell everywhere.

Lifeforms divided into the quick and the dead. Millions of creatures died in the first few minutes, drowned, burned, crushed, suffocated. Millions upon millions followed in the next days and weeks and months as ecosystems changed too radically for them to keep up. Seventy percent of all living species died out. Many regions were simply washed clean of life. After 150 million years of dominance, the dinosaurs ended their reign on earth, leaving their only descendants, the birds.

Something like this event may have occurred about sixty-five million years ago. Roughly five hundred years ago, something similar happened to scribes when an obscure goldsmith named Johann Gutenberg invented his printing device. For them, the change was almost as dramatic and, in the long run, just as deadly.

On the Shadows of Ideas

Life for a European scribe in the early fourteenth century was pretty settled. He (almost all were male) was either a monk slaving laboriously in a monastery or a copyist slaving laboriously in a sweatshop. He knew that his ancient and honorable life-style would never change. There was always demand from the few lords for hand-lettered books and from the few literate burghers for religious tracts. There were also the thirty European universities and their constant need for new manuscript copies. Still, the scribe's mainstay was the church, which ordered thousands of copies of papal indulgences to sell to the laity, and the business community, which needed simple contracts to conduct its growing long-distance trade. Life was, if not good, at least very stable.

Then, in the middle of the fourteenth century, the Black Death struck, killing between four and six of every ten Europeans—perhaps forty million people. Whole villages were wiped out, vanishing utterly from the face of the earth. The plague returned in waves from then until well into the fifteenth century, devouring lives like corn stalks before the scythe.

By the middle of the fifteenth century, a scribe's life had changed. There were now only half as many Europeans as there had been a century before, but the number of people who could read and write, always a tiny minority, was decimated. In those days, most literate people lived in cities, and the dense, squalid cities were the hardest hit by the plague.

All over Europe, there were surplus goods and empty lands; with half the population dead, the survivors took possession of their goods. The economy took off. The few scribes, clerks, and notaries who remained were busily scribbling away, trying to supply an ever-expanding demand for documents.

Then, Gutenberg's newly invented printing press, by making printed copies cheap relative to handwritten copies, drastically changed everything. By the end of the fifteenth century, his device was taking over the manuscript market, throwing scribes out of work and explosively increased the number of available books. By making book copying much cheaper, it made books cheaper and more common. That, in turn, led to the eighth wonder of the world, that new thing—the bookshop.

Before print there were only a few thousand manuscripts in total. But by the early sixteenth century, twenty million books had been printed—almost as many books as there are in the U.S. Library of Congress today. Books stopped being rare and expensive objects of art and religious meditation or the secret codices of a guild, church, or government. The job of scribe was starting its century-long slide into oblivion.

Eliminating scribes, however, was only a minor part of the print revolution. Printing led to page numbering, indices, and bibliographies, which were now possible and made searching easier. This development forced people to learn the alphabet so they could use the new indices. The new printed books were cheaper, more widely distributed, more accurate, more portable, and more convenient than manuscripts. They in turn led, eventually, to increased literacy, standardized spelling, and heightened author accuracy. They created libraries and even the idea of authorship; for the writer was no longer just an anonymous copyist and so could be held accountable for the book's ideas.

Printing brought about enormous changes in medieval Europe. For example, because it began to standardize spelling and was relatively cheap, it solidified the various vernacular languages. Before its invention, there was no such thing as standard English, or French, or German. Many European countries were a gaggle of squabbling language groups and the only real exchange of knowledge came through Latin—the common language of the literate world. So, by making the distribution of knowledge cheaper, printing standardized English, French, German, and so forth.

The tidal wave of new books spread and democratized knowledge. The tiny elite of priests, lawyers, aristocrats, and guildsmen could no longer control Europe's few precious bits of knowledge, whether in metallurgy or liturgy, astronomy or gastronomy, philosophy or choreography. In particular, the new books led to the first printed tracts challenging the church's dominance in religious knowledge. By the 1520s, the beginnings of the Reformation and the birth of Protestantism followed.

Printing also led to other deep and subtle social changes. It was the television of its time. For example, it decreased the importance of memory and its main possessors, the elders, thereby devaluing the importance of tradition in council chambers across the land. That decline in turn added fuel to the humanist movement and spurred on the Renaissance.

Education, engineering, science, and technology transfer—all profited from the spread of printing, which contributed to both the gathering of knowledge and its inevitable splintering into new disciplines ruled over by a few widely separated experts. It turned the pursuit of knowledge into something almost anyone could participate in. And it created publishers.

When inventing movable type, Gutenberg wasn't thinking of the revolutions that would follow it. He certainly didn't anticipate Adam Smith's *Wealth of Nations,* Charles Darwin's *Origin of Species,* or Adolf Hitler's *Mein Kampf.* Nonetheless, his device changed everything, and it changed everything forever.

The printing presses destroyed the scriptoria, drove out the jealous scribes, and released knowledge to run free throughout the world. With printed books acting as vectors, the new disease of cheap and easily communicable knowledge spread all across Europe; the speed of change accelerated as ever larger numbers of people were infected with ideas they would never have met with when books were rare and expensive. As a consequence, the pace of technological change picked up.

Whenever we improve the production, handling, and distribution of information we drop the price of thinking. This always has big consequences. Although we don't know the details, we do know there must have been a major leap forward when we invented language perhaps half a million years ago. Similar breakthroughs occurred when we developed clay tablets, perhaps six thousand years ago; papyrus, maybe four thousand years ago; parchment—and then paper—roughly two thousand years ago; and, finally, movable type, five hundred years ago. Now it's the computer's turn. Massive technological change is once again coming to the one technology we've had throughout recorded history—the same technology we use to record that history—writing.

Working in Clay

Over the past fifty years, printing, paper, and transportation costs have all risen, while the cost of their electronic counterparts—computing, electronic memory, and telecommunications—have roughly halved every two years. Now the two cost curves have crossed, and it's cheaper to make and distribute electronic books than it is to print paper ones. The only thing holding electronic books back is the lack of cheap, sturdy, portable, high-definition electronic displays. Within a decade that will change. After five hundred years of business as usual, publishing is once again about to change—drastically.

Of course, paper books still have some pluses. They are hard to break. They don't need batteries. They're readable in the bathtub and in strong sunlight. They've been around for over five hundred years, and all literates know how to use them.

But they have minuses, too. Illiterates can't use them. They don't talk to us, adapt to us, or have animated illustrations or music. They don't let us change font size at will or zoom or pan illustrations; nor do they recognize voice commands or visual cues. They make it hard to combine the information in different parts of one book or in many books on related subjects. They aren't easy to turn into Braille or big print. They aren't cheap, long lasting, or easy to copy, obtain, or search. They certainly are not portable in bulk.

Electronic books could be all these things and yet retain many of the pluses of paper books. We could handwrite marginal notes, highlight them with colored markers, and put bookmarks in them. Cross-referencing could be either reader-controlled or computer-generated. Books could be customized by—or for—their readers; each copy need not be exactly the same.

Electronic books could also let authors mix voice, music, color, video, pictures, numbers, and text. Today's automated book-reading machines can even read the text of any printed book out loud—a boon to the blind, disabled, illiterate, or busy.

Perhaps a single electronic book might, in the future, become a whole library of children's books. The book could listen to the parent (or child) reading aloud for a few hours until it can read any of its repertoire of books in that person's voice. Imagine children's books that can read themselves to a child at bedtime. By listening to the child's breathing, the book could reduce its volume, dim the lights, and slow its cadence as the child drops off to sleep. Of course, it can't kiss the child goodnight. But then, neither can a paper book.

In sum, about all we can really say in paper's favor is that it's lighter than clay, less awkward than papyrus, and cheaper than parchment.

Thinking Machines

It is a curious fact that in America, a supposedly highly literate nation, a hardcover book is one of the year's top twenty-five bestsellers if it manages to sell only 115 thousand copies—enough for about one-twentieth of 1 percent of the population. *Gone With the Wind* has sold 21 million copies over forty years, but 55 million people saw the first half of the movie on television in one evening. *Roots* has sold 5 million copies over eight years, but 130 million people watched the television version.

Roughly 27 million American adults are functionally illiterate—about one in every five. Every year almost three-quarters of a million American high school students drop out, while another three-quarters of a million graduate unable to read. In addition, the percentage of students graduating from high school has decreased every year since 1984. The social problems causing the dropout are serious, and most are unrelated to books; yet, while electronic books are obviously no cure-all, they may help to reverse the trend away from social involvement. Since talking books de-emphasize literacy, they may also help enfranchise the illiterate, the dyslexic, the blind, the disabled, the elderly, and the young.

We have the books we have today, not because they're in the best possible form they can be in, but because paper technology limits what they can be. It need no longer do so. For example, imagine learning orbital mechanics through a video game that lets us choose burn rate and burn time, then shows us what happens to the rocket. Once our interest is captured, the book could explain more of the physics involved. Or imagine a chemistry book that lets us bring together different molecules and watch what the atomic forces do to them, following through until the molecules reach a stable state. What about a biology book that takes us inside a working cell, letting us see the cell in operation and showing us what each part does under normal or disease conditions?

Perhaps a future mathematics book will let us choose our own parameters for mathematical functions, displaying what happens to their derivatives or, even, dispensing with simplistic calculus models entirely let us work directly with simulations. Imagine a statistics book that dispenses with artificial measures like averages and standard deviations and gives us the raw information and lets us interact with it. Imagine a physics book with an animated Galileo, Newton, and Einstein to explain their various theories, then guide us through the consequences, letting us ask questions or suggest alternatives. (As computer technology improves, we might change our guide to whomever we wish: perhaps a favorite aunt, Bugs Bunny, or Queen Elizabeth II.)

Imagine a computer book that lets us tour a computer chip. The book first displays a chip as we normally see it—a black fleck of shiny silicon. The book has two controls: a joystick and a light-dimmer switch. As we move the joystick, the book displays the image we would see if we were at the chosen distance and point of view. The dimmer switch controls the time scale; twisting it changes the speed at which things happen.

Pressing down on the joystick brings up a quarter-scale display in the lower right-hand corner with text, voice, or a video of the author explaining what we're seeing and pointing out other interesting things we might like to see. As our viewpoint approaches the surface of the chip, it expands to cover the entire display; then the horizon disappears off the screen entirely. As we move in we see pulsating rivers of light representing electron flows, and we hear a whispery rustle representing thermal noise, which grows to a keening roar as we get nearer one of the rivers of light.

At even closer range, and further reducing the time scale, we see individual clumps of electrons switching through individual gates. The sound also slows, so we can hear each electron whizzing by. Zooming in yet closer we see a single electron about to tunnel out of a channel.

Pricier versions of these electronic books could replace the joystick and dimmer switch with touch-sensitive screens and simple vocal commands: *stop, go, faster, slower, zoom here, pan around, what is that, show me more, tell me why.*

And what works for technical subjects can work for anything else. For example, imagine an electronic atlas that opens with a rotating globe. We learn about different parts of the globe by touching it. We can then find out about the geography, history, politics, culture, or economics of each area. Touching economics brings up overlays showing trading partners, trade routes, and goods. Touching any trade good, from tractors to video cameras, gives us information on the source of all the raw materials used to make it.

Activating another portion of the display tells us something about a region's history, geology, demographics, transportation systems, climatology, political allies, nearness to major fault lines, chlorofluorocarbon emission rate and projected five-year development, skin cancer rate over the past decade and next-decade projections assuming various levels of ozone depletion.

Another portion of the globe might let us model and extrapolate land use and deforestation rates over time to examine the effect of different tariff levels; or measure the effect of waste heat from cities on fish populations, of power lines on bird migratory paths, or of global warming on coastlines and industries.

Touching yet another portion gives us the region's Nobel prize winners—with vocal explanations of their achievements—or displays photos of local politicians. We could look at a breakdown of the region's gross product, along with government budgetary expenditures and fiscal projections for the next year. Perhaps we want to know about the effect of solar flares on the region's satellite reconnaissance, or the region's offshore natural gas deposits, or the epidemiology of retroviral disease. Whatever our interest, all portions of the display could be in color and could be accompanied by movie snippets, stills, voiceovers, and music. For us, the world would become, literally, an open book.

We could have tourbooks for trees or television sets, space shuttles or lungs, solar systems or whales. A computer book could let us build a model of a computer and run it, rather than laboriously trying to teach us how current computers work. A business book could make us president of a corporation, then teach us about the business practices we need as we try to run the company. A classical music composition book could make us a conductor or concertmaster or second chair flute, simulating the music from other orchestra members as we learn to establish the tempo and interpretation.

When students have such books the role of teachers will change. Reading one of these books on, say, cosmology, isn't like studying cosmology; it's more like becoming a cosmologist. Think what an adventure interacting with such books would be. We would feel more empowered, more in control of our own learning, because we would be building models of things within the book itself. We would be learning for ourselves, not being talked at.

The Frailties of Print

If you publically ask publishers what they do they usually reply with uplifting homilies about disseminating knowledge and protecting intellectual freedom. Privately, though, they say they are booksellers. And by *books* they mean bound woodpulp products smeared with petroleum residues. That's so because when books are printed on paper there's no essential difference between a physical book and the information in it.

That needn't be true for electronic books. The display is one thing; the information displayed is another. We could show many things on one display, or exhibit the same thing on many displays. The display becomes a window onto another world through which we can see whatever we wish to see. A single electronic book could thus be more accurate, flexible, informative, sophisticated, and adaptable than any number of paper books. Because, unlike paper books, electronic books could be connected directly to the world's information sources; they could keep themselves constantly up to date.

Paper books can't talk to libraries and other information sources; so today's authors have to first think up possible questions, research the answers, then find ways to summarize them in print. With electronic books, the reader could pose the questions—questions the author may not even have thought of—the book could research the answers—perhaps almost as well as the author—and let the reader decide how best to display the information. Such books might increase comprehension, retention, and emotional response by combining the best aspects of lectures, television, computers, and print.

Of course, they probably won't. There will still be good books and bad books—perhaps in the same proportion. Electronic books might even be much worse. Because of their great immediacy, they could shape our unconscious responses more deeply. So bad electronic books could be much more dangerous than printed ones, just as a demagogue's speech is more compelling than the text of the speech.

Reading is work, but long before writing we had speech, sounds, and sights. That's why television is so much more popular than books. We've only had about five thousand years to get used to writing, while we've had millions of years in which to hone our audiovisual responses. So we're good at interpreting sounds and sights—particularly if we're in control and can stop, replay, or interact with the action at any time.

We could have interactive books on subjects ranging from foreign languages to painting, from piano playing to architecture, from cooking to hairstyling. Such books could change the way we think, the way we work, and the way we see ourselves, our artifacts, our societies, and our world. With books like those we'd be exploring, not reading.

With such books we need no longer be intimidated by unnecessary formality, nor by the artificial linearity authors are forced to place on a subject just to fit it into the unnatural confines of paper. The difference between those books and paper books could be the difference between behavior and the description of behavior. They needn't simply talk about what could be done; they can show us.

The Geography of Information

None of the changes books are going through today mean that publishers will go away. Currently, about a thousand books a day are published worldwide in all languages. As that number continues to mushroom, publishers can only become more important, not less so. Like librarians, booksellers, and reviewers, they function as stamps of approval, arrangers, and labelers. Without them, or something like them, we would simply drown in information. We would have to spend too much time trying to tell the good from the bad, the useful from the irrelevant, the interesting from the boring, the erotic from the trite.

In the future, however, publishers and librarians may not be the only ones who do these things. With the explosive growth in electronic information, a whole new profession may develop—people who find things— perhaps they'll be called *ferrets*. For those who want to rummage for themselves, there may be another new profession—people who organize things. Maybe they'll be called *mapmakers*. And everyone will need people who select things, distinguishing the good from the bad; perhaps they'll be called *filters*.

These three professions mirror the three basic aids currently found in nonfiction books: indices (ferrets), contents pages (mapmakers), and bibliographies (filters). They also correspond to the three basic uses for computers: searching (ferreting), sorting (mapmaking), and selecting (filtering). These are the things that publishers really do, despite their present belief that they're distributors of woodpulp and petroleum.

In the future, your computer may run hundreds of ferret programs, continuously exploring the world's information for useful tidbits. When a ferret returns it may have to face dozens of filters trying to prevent it from adding all the information it found to your personal information base. Information that the filters judge relevant to your interests (and that costs a reasonable amount) is passed to the mapmaker, who links it into your personal map of what's important.

Businesses are more than ready to pay for such precious information. In most corporations today, middle managers play the part of ferrets, mapmakers, and filters for senior managers. The paper reports they often produce, however, are hard to search, index, and compare. Once we commit something to paper it's frozen. We can't display it in other ways—say with histograms, pie charts, or graphs. A table listing national populations alphabetically by country is very annoying when we only want to know the top five populations. The information we seek is there, but it's hard to get at.

Further, the economics and bulk of paper restricts the usefulness of books. *Compton's Encyclopedia,* for example, now comes on a computer disk that contains almost nine million words; over five thousand articles; almost sixteen thousand photographs, maps, and diagrams; an hour of recorded voices and other sounds; forty-five animated sequences; *Webster's Collegiate Dictionary* (itself with sixty-five thousand entries); and a word processor. There's no way the publishers could get all that information on paper.

Look, for example, at today's Yellow Pages. To get the most benefit from the book we must understand exactly how the telephone company organized it. We must also have a detailed map, a subway guide, bus routes and schedules, the local Better Business Bureau report, and plenty of time. The phone companies give us no choice because a better Yellow Pages would be far too expensive, bulky, and slow to produce in a paper format. If we do it electronically, however, we can store and update a lot of the same information cheaply and quickly.

To be really useful, Yellow Pages should list all businesses by each street, neighborhood, and mall; by the time needed to get to them from our current location; by whether they're in a safe neighborhood; by their nearness to various landmarks; by whether they're currently having a sale; by whether they accept checks, cash, or credit cards; by their hours of operation; by their nearness to restaurants, gas stations, public restrooms, malls, or anything else we might care about; by their costliness, reliability, revenues, experience, and returns policy; and, finally, by what past customers say about them. The same thing applies to every other kind of business or professional: from doctors to car mechanics, from lawyers to window cleaners. All businesses would pay Yellow Pages to be included in this sort of directory, for the same reason that they presently pay credit card companies—because it could mean more business. Today only 5 percent of the roughly 9.5 million American businesses advertise outside the Yellow Pages.

In thousands of similar ways, paper technology severely limits what we can do, what we can know, and how fast we can learn it. To most Americans, for example, the twenty-five million books in the U.S. Library of Congress, perhaps the nation's greatest intellectual resource, are less useful than a cheap home encyclopedia.

Dancing in the Light

Electronic books have already become a part of the book market. Yet some publishers still resist them because making their books electronic makes it harder to ensure that they aren't illegally copied.

Traditionally, publishers and authors have used copyright and the courts to protect themselves. So it would be only natural for publishers to adapt to the new electronic technology by trying to copy-protect their books. Copy protection is like putting a lock on each copy, then selling a key with each locked book. Both software and video producers tried— and recording artists are still trying—to protect their products in this way. If you copy-protect your product, they reasoned, you can charge whatever you want for each copy. It sounds great. But it didn't work.

Intellectual properties like books and movies and songs aren't the same as tangible properties—like ham sandwiches—or rights like franchises, licenses, water rights, stock futures, or airline routes. Often, copy protection merely feeds lawyers and annoys valid users; certainly it adds expense and works against easy searching and comparison. But it seems unavoidable. Or is it?

The information in books is freely accessible; that ease of information exchange makes modern civilization go. Yet paper technical books aren't easy for readers to search, cross-reference, reindex, or carry around in bulk. Suppose, for instance, that you're reading this book on paper and want, right now, to find all other books like it, compare it with ten others, talk to others who've read it, or find discussions of similar books. Well, you can't. At least, not easily—or cheaply.

So if there were some way for publishers to make their books cheap, electronic, and unprotected, it would increase information production, comparison, linkage, and dissemination. Which would benefit us all. The problem, however, is copying.

The Information Universe

The Xerox Corporation estimates that two thousand million pages are currently copied every day worldwide. Fast and portable color copiers are now so cheap that enforceable copyright is rapidly becoming a thing of the past. Publishers can sue big photocopy stores, but they can't sue half a million homeowners. And the copying problem can only grow worse. Copiers armed with capacious electronic memories are getting smaller and cheaper and faster year by year.

But the copying problem simply explodes as books go electronic. Copying electronic information is easier, cheaper, and faster than copying paper and can be done without human labor. When books are electronic, even if they are encrypted, at some point they must be decrypted for the user to read. At that point they can be copied. Perfectly. Further, the cost of that copy, which is equal to the cost of the memory needed to hold it, is quickly approaching zero. Even if publishers try to avoid electronic copying by staying with paper, readers could electronically scan their paper books, then copy the electronic copy. It will merely take longer to make the first copy. So if books continue to be priced much higher than the cost of copying them, publishers and authors will inevitably lose serious money to pirates. Which will harm all of us.

Information isn't the same as tangible goods. It can be copied almost instantly over enormous distances, leaving no trace of the copier and suffering no loss in fidelity. When a few million of us have the means of duplication in our hands, copyright may still exist as a legal idea, but it will be unenforceable between publishers and the public. Today, no one is arrested for privately copying and sharing audio, video, and computer tapes and disks, even though that practice has never been tested in court. So, to survive, publishers will probably have to make book buying roughly as cheap and as convenient as book copying.

If that sounds ridiculous, then perhaps you're thinking of a book as a physical object that costs a lot to make and distribute new copies. But with electronic books the cost is in the production of the book, not in its reproduction and dissemination. Compared to paper, copying and transmitting electronic information is essentially free to the publisher. If publishers try to charge high per-copy prices, some customers may use the same technology to make their own essentially free copies. So the publishing system must change.

Yet, if authors and publishers can't protect their books, they can't possibly profit. If they can't profit, how will new books get made? To find the answer to that particular puzzle let's suppose we live in a universe where there is a magic machine that can copy anything—a book, a chair, a computer—quickly and cheaply. We could buy a sofa, some sandwiches, and some silverware, then copy them at home. The food, forks, and furniture markets would then evaporate. In essence, that's the universe of information. In that universe it's as easy to copy something as it is to produce it. All the cost lies in the design, not in the manufacture or reproduction. It would then be foolish to try to sell food, but we could still sell recipes and—more to the point—the rights to future delicious recipes.

Subscribing to the Idea

Many early software companies hired large numbers of programmers. To support a big payroll (and sometimes through just plain greed), they charged high prices for each copy of their software. Which led to piracy; which led to copy protection and even higher prices; which led to even more piracy. Software houses eventually broke that cycle by abandoning copy protection altogether. They realized, in effect, that they could make more money by selling rights to future food rather than food itself. So they hooked their audience with good, cheap products and a promise of continuous updates for a yearly fee. It worked.

Unlike those early software companies, most of today's publishers don't hire authors. Instead, they encourage free-lance authors to write books, then help develop the projects and promote and sell them. Publishers provide the capital and expertise to develop titles—getting titles from supply to demand—and get money from demand to reinvest in supply. Publishers, or something like them, are necessary.

Electronic publishers, on the other hand, don't necessarily have to sell their wares the way paper publishers do today. Like selling magazine or cable subscriptions, having a large and stable number of customers each paying a small amount per day is less risky and more profitable than having a few incidental buyers of expensive single copies. The uncertainty caused by focusing on selling single copies is what's wrong with publishing as a business today. Books could be more easily distributed if they were electronic, and publishers could profit without copy protection. Subscription could make books cheaper for both publishers and readers, reduce the risks of publishing, and increase profits. It would work by shifting publishing's emphasis from betting that one particular title will be a bestseller to maintaining many readers of at least one title.

In a decade or two, publishers will be back to doing what they do today. Once the novelty of electronic books wears off and everyone is offering them, publishers will again compete mostly on the basis of their books' design, packaging, and promotion. In the meantime, however, publishers unprepared for change in the book industry are in for some fun.

The Electronic Bookstore

Paper publishing is risky business. The economics of printing forces publishers to produce titles in large printings. Because per-copy costs drop sharply with volume, small print runs are not profitable. Large print runs, however, mean that more capital is tied up in paper for as long as the copies take to sell—if they ever do. So less capital is available to buy new titles or to promote current ones. In the meantime, the costs of warehousing, security, and insurance pile up.

Even when publishers predict demand accurately, printing alone adds four to six weeks to delivery time and mailing adds a further two to three weeks. Even if warehousing and capital costs were zero, storage would remain a problem, for paper can't be kept waiting indefinitely because it decays in a few years. Besides, since paper books can't update themselves, they and their dated information are soon obsolete.

Because of those constraints, publishing today proceeds by guess and by gosh. Many more books are published than there is retail space for, and few of us buy books anyway. So a publisher lets retailers return unsold copies to increase the chances that they can afford to carry new titles. Occasionally as much as half of a mass market fiction print run of half a million copies is returned and destroyed. With the ever-rising tide of new books, the average newsstand display time for a title is now around a month. Books are, in effect, becoming magazines.

With electronic distribution, however, retail outlets won't have to keep as many copies of each title on hand as they think they can sell. They will need only one. Printing copies in the store or delivering them over the telephone will dramatically increase the diversity of titles that booksellers can offer.

For publishers, going to electronic books could mean no printing and its costly consequences: warehousing, transportation, delay, backordering, remaindering or destroying unsold books, and losing business. Instead of killing trees, expending oil and human labor to transport books, and polluting the environment, publishers could send any book on demand directly to any reader in the world in minutes. The same system would work for any other form of information—movies, software, designs, music, pictures, and television or radio shows.

By century's end, a few bookstores may devote half their space to books on computer disks, thereby at least quadrupling the number of titles they can carry but otherwise keeping many of the problems of paper publishing. Retailers will still have to order as many copies of each title as they think they can sell. But by early in the next century, some bookstores may become wall-sized, touch-sensitive display screens electronically displaying an array of titles, along with pictures and commentary. Perhaps each title will have its own book-sized rectangle on the display. Customers buying copies would read them on their own portable book readers.

Book chains would love such displays. They would have low rents, low insurance costs, no staff, no stock, no warehousing, no transportation, no remainders, no returns, no overhead, no reshelving costs, no replacement costs, no shoplifting, and no health care or workers' compensation expenses. The display would be easy to reorganize and could operate twenty-four hours a day. Finally, it could be placed at bus stops; in laundromats or church yards; or on playgrounds, buses, trains, ships, and even planes.

When books are electronic, we will have instant and unsleeping access to them—no more worrying that the library or bookstore isn't open late. Also, we'll have instant updates and revisions based on the latest information and electronic contact with all other readers of each book, thereby sharing ideas and reactions more rapidly and with more people.

Electronic books need never go out of print, and they'll be cheaper and far less bulky than paper books. Already we can store thousands of books on one small, lightweight computer disk at less than a cent a book. That price will continue to halve every two years. Paper simply can't compete.

Electronic libraries won't need big and expensive stores of decaying paper; so they will be able to shrink from warehouses to rooms and, one day, to a single telephone or cable television line. All catalogs will be electronic—as many already are—making them easier, faster, and cheaper to search, produce, and update. They could make it easier for libraries to refer readers to other books on similar subjects, tastes, or interests.

Libraries won't need to rebind old books or buy multiple book copies to allow for scuffing, mutilation, and destruction. Binding magazines into volumes, reshelving books, and reserving and otherwise restricting access to them will be things of the past. Librarians will no longer live in mortal terror of fire and flood. They won't need to chemically treat their decaying books, microfilm them, or transcribe them by hand to Braille, big-print, or audio formats. Currently, the U.S. Library of Congress can afford to transcribe only two thousand new books and one thousand new periodicals a year. Out of its twenty-five million books, it carries only thirty thousand in alternative formats.

Of course, some of us will always dislike having to read books on a screen. But once a book exists in electronic form, it is easy to print it out on paper. Besides, display technology will surely continue to improve so that in a decade it will be nearly as good as paper. A decade after that, it will be better.

Even if paper books persist, their appeal will fade. As electronic books quickly outgrow the restrictive linear text-and-pictures format, readers will experience paper books as flat, lifeless versions of what are now, essentially, living books. Going electronic is the future of books.

Getting There from Here

Some publishers will probably fight rather than switch. They may protest the new technology and push for laws against copying or for a legally enforced standard electronic book display that tries to keep book production or book copying out of our hands. It is easy to predict that because that's exactly what music, software, television, and movie producers each tried to do when cheap copying technology first entered their industries. Movie producers, for example, initially thought the videocassette recorder would destroy their business. Today, they can't live without its income. In America alone four thousand million videotapes a year are produced.

Fifteenth-century scribes and eighteenth-century weavers tried the same resistance tactics when their ways of life were invaded by machines. Today some military pilots are fighting a similar war against the computer's incursion into the cockpit. Successful people in every field usually resist change—it might, after all, be for the worse. Nonetheless, such publishers' efforts will eventually fail, just as those earlier attempts failed. Paper books are a flawed and expensive technology, and copy-protecting electronic books is a bad idea in the long run. Do we weep because pocket calculators wiped out the slide rule industry? Or that polio vaccines destroyed the iron lung industry?

From a standing start in 1984, compact discs overtook phonograph records in just five years. Paper books will take far longer to die, partly because of inertia, but mostly because cheap, sturdy, portable, high-quality displays aren't here yet. But it's inevitable. Computer technology usually halves in price roughly every year-and-a-half to two years; at that rate, good displays will be thirty times cheaper in seven to ten years. In fifteen to twenty years they'll be a thousand times cheaper. So two decades from now paper books may be the phonograph records of their time. They'll exist for historical, sentimental, or ceremonial reasons. And for the wealthy, they'll still make perfectly good furniture. But eventually they'll go the way of the vacuum tube, which, legend tells us, existed in the 1940s and 1950s.

In advanced countries today, those thirty and under have no special interest in paper technology. They're more familiar with television and computer screens than with paper books and newspapers. By 1996, they have had Pac-Man for fifteen years, MTV for seventeen, Apple Computer for nineteen, and *Sesame Street* for twenty-seven. Americans alone have over thirty million personal computers and more than thirty million video-game machines in their homes. Over 70 percent of all American homes with a child between eight and twelve have a Nintendo machine. More than 95 percent of all American high schools have computers. Total sales of personal computers have exceeded a hundred million, with sixteen more million to be added in 1996. The U.S. Department of Commerce estimates that a third of all American three-year-olds have already used a computer. Today's generation talks about computers the way their parents did about cars—or drugs. Their first instinct usually isn't to reach for a paper book.

Books on the Margin

If future publishers charge roughly as much as it costs to copy a book they won't have serious losses from piracy. It would probably even be reasonable to charge, say, five times the copying costs, since they would be the most convenient—and legal—suppliers. But prices over, say, ten times the cost of copying may lead to a good deal of piracy, and the potential pirates aren't just a few villains twirling their mustaches in dark electronic corridors of the international computer network. They are also millions of people sharing favorite books and magazines with friends.

In an ideal world, then, each electronic book copy should cost between fifty cents and a dollar. That sounds crazy when we look at today's five-dollar paperbacks and twenty-five-dollar hardbacks. But let's do the arithmetic.

Suppose that a company publishes only two hundred titles and that each title finds only three thousand readers; the publisher then has six hundred thousand readers in total. Imagine that only one in ten of these are serious readers and wish to subscribe. At ten dollars a month for the subscription (that is, for the subscriber's right to pay one dollar each for each copy of any book they want), the publisher would gross over seven million dollars a year.

That's not counting income from the seemingly paltry one-dollar-a-copy charge. If the average subscriber only buys five books a year, that alone comes to more than a quarter million extra. Even if subscription halves in a year, the publisher will still gross over three million that year. Further, sizable chunks of today's costs for capital, warehousing, transportation, and so on would be replaced by far smaller electronic costs. So the net profit should be higher—and the net risk lower. All that with only two hundred low-demand titles.

Of course, this calculation will only be good for the next five years or so. Once many large publishers are offering such services, there will be more subscription services than the tiny reading market can bear. So eventually, perhaps over the next ten to fifteen years, publishers might become the equivalent of common carriers, and a new branch of specialized publishing might offer subscription services in specific areas— hunting, computing, movie books, and so on—by pooling small parts of the lists of several publishers.

When books are electronic, the marginal cost to distribute a copy of any title will be near zero. So it will be irrelevant that some titles aren't big sellers—they cost nothing to keep, they can be stored forever, and they cost almost nothing to distribute if demand ever rises. These economics will shift the publishing company's interest from single titles to its entire list, for each title will be valuable if it eventually gains the publisher more subscribers. Even snob appeal books that no one ever reads will be useful if they add luster to the list and so attract subscribers. Publishers could support more authors and a wider variety of specialty topics, and there could be more books. They could always be up to date and bigger, cheaper, and easier to get and use than present-day books. And they could always be in print.

The returns from expanding the number of distribution sites is even more dramatically nonlinear than increasing a book's print run today. Each distribution site could be simply a telephone line and a small special purpose computer—a few thousand dollars of equipment that would easily pay for itself even if it only supported a few dozen new subscribers. Under these circumstances, anyone could be a publisher. A few thousand dollars worth of computer equipment could translate into millions of dollars in subscriptions. Rents too could be low as distribution sites needn't be in Manhattan or London. As no employees are necessary, Nome or the Outer Hebrides would do just as well.

More numerous publishers would increase title diversity, leaving the market alone to decide which ones are good. Today, because we commit everything to paper, the few can control what the many can read by controlling the bottleneck—the printing process. This is about as sensible as letting Kodak control the movie industry because it produces the most film.

Further, electronic publishers could be international without having to staff and support overseas bases. Electronic information, unlike paper books, is easy and cheap to transport, so distance doesn't matter. Geography, at least on planet earth, will be nearly irrelevant in the information universe.

At that point, our sluggard governments will start worrying about export controls and tariffs—but they too will be irrelevant, because unenforceable. The state can no longer control its population by restricting access. Cutting off international calls, for example, would have drastic economic, social, and political consequences. And even if a state were to disallow international calls, there would still be satellites and radio. That's as true of Russia, Singapore, and China as it is of America, Canada, and Britain.

The Quick and the Dead

[I]t at once struck me that under these circumstances favourable variations would tend to be preserved and unfavourable ones to be destroyed. . . . Here, then, I had at last got a theory by which to work.
Charles Darwin, *The Autobiography of Charles Darwin*

Few people will risk jail to copy and try to sell a product that anyone anywhere in the world can legally and almost instantly get (plus continuous updates) for roughly thirty-five cents a day. When any one copy of a book has a marginal price of about a dollar, every book can always be in print and distribution can be cheap, immediate, and widespread. In such a world, both pirates and used-book stores will cease to exist, and publishers won't need to copy-protect their books.

Publishers could then tailor their books to specific groups of subscribers, based on information that subscribers volunteer about their tastes and interests. Eventually publishers could even tailor books to individual subscribers, thus becoming immune to the raids of pirates producing uniform copies of one version of each book. Nobody would buy off-the-shelf clothes if tailor-made ones were just as cheap and available.

Yet, while interesting, a future full of electronic books won't be a simple world. Eventually the communication, computation, and entertainment heavyweights will weigh in and crush most of the small fry. Even that, of course, might be good in the long run, since to do so they'll have to use economies of scale to bring the price of display technology low enough for the general public to afford it; and they'll have to promote it heavily to develop a big enough market for them to swim in. Once that first tidal wave passes, however, small independents will start up again to service various niche markets.

To put all that change in perspective, it might be helpful to look at Western Europe in the middle years of the fifteenth century, just before and during the early days of printing. The most virulent outbreaks of the Black Death had just passed, and the economy was beginning to recover. Chaucer had been dead for fifty years, Columbus was an infant, and Shakespeare wouldn't be born for another hundred years. England, Scotland, Ireland, and Wales were still independent countries. France, just recovering from yet another long and pointless war with England, was a set of feudal duchies and principalities. Germany didn't exist yet; it was still a set of warring princes rattling around in a power vacuum. Italy was fragmented politically. One of its independent city-states—Venice— thanks to its stranglehold on Asian trade, was the single richest European power. Spain was simply four kingdoms—three Christian, one Islamic. Constantinople had just fallen to the Turks, and all Europe seemed sure to soon follow. Most medieval Europeans viewed the future with despair, thinking that their best days were behind them.

This was the setting into which printing emerged in the 1450s and everything immediately started to change. Living in a dark time, Europeans used the new books to spread the fire of knowledge everywhere, and that fire nursed along a Renaissance, a rebirth. Everywhere, new attitudes and new inquiries took shape and blossomed. Everywhere, ancient authority and ancient custom were questioned.

The changes were not solely the result of printing of course. But print played a large part by improving the speed of creation and the dissemination of knowledge, which always lead to big changes in how we live and what we think.

Still, every change has a potential negative side; that's why so many of us fear it. The bacteria that caused the Black Death, for example, had been living in rodents in central Asia for who knows how long. They were introduced into Europe only when Europeans started to form permanent long-distance trade links with Asia. Now they're all over the world. No matter where you live, if you camp out or live with a pet near a wooded area anywhere on earth, you too could catch it.

Change is frightening, because it can have unknown side effects. But it's also inevitable. Established publishers should brace themselves for heavy weather ahead. The world they're used to, the world they've grown comfortable in for the past five hundred years, is about to change again. Once again life-forms are about to be divided into the quick and the dead. The dinosaur killer has already descended through the atmosphere. And now it's about to hit.

4

Only Connect

Cheap printing, wireless telephones, trains, motor cars, gramophones and all the rest are making it possible to consolidate tribes, not of a few thousands, but of millions.

Aldous Huxley, *Those Barren Leaves*

Thought is subversive and revolutionary, destructive and terrible; thought is merciless to privilege, established institutions, and comfortable habit. Thought looks into the pit of hell and is not afraid. Thought is great and swift and free, the light of the world, and the chief glory of man.

Bertrand Russell

The tablecloth is snowy white, recently cleaned with a new product designed by a laboratory in Switzerland and made in Connecticut. The table is decorated with a vase of flowers flown in that morning from Israel. The heavy cutlery is from Germany. The cut glass decanters are from Ireland. Your waiter brings you the wine list as you sit chatting with your friends. The sun, a sinking red ball, is painting a flight of pigeons with earth tones of red and gold.

You're in a trendy Manhattan restaurant about to have a meal in the global village. You order a French wine and an Italian pasta. The wine was flown in on planes owned jointly by an American and a British firm. The wine is from Provence, but the pasta isn't Italian, although its recipe is. Your friends order food mostly grown nearby; but the chilies are from Mexico, the curry from Pakistan, the fresh fruit from Venezuela, the fish from Alaska, the grapes from California, and the cheese from Canada.

Even the food grown in Iowa or Indiana or Illinois was processed through a complex system of trucking, insuring, legal, and retail firms before arriving on your plate. And it probably had to go on trucks or railway cars made, or partly made, in Japan, Britain, France, or Germany.

While you're chatting, Jodi, one of your friends, starts to beep, excuses herself, and reaches into her jacket pocket for her telephone. She's talking to a business associate, a mutual friend in Curaçao, about a shipping deal involving partners in Britain, Singapore, and Hong Kong.

As she speaks, the computer in her telephone is turning her voice into a series of pips and tossing them into the air. A radio receiver several blocks away picks up the low-powered radio waves, uses its computer to check that it's an authorized call from that telephone, amplifies the signals, and sends them on to a nearby central station operated by her cellular-phone company. The station decodes the signals, creates a corresponding pulsetrain of microwaves, and beams them up to one of hundreds of orbiting satellites. Computers in the targeted satellite decipher the call's destination and beam microwave signals down to a receiving station in Curaçao. The station then routes the signals to a portable telephone held by your mutual friend as he speeds down a rutted highway in an American car. All the various encodings, decodings, routings, reroutings, uplinks, and downlinks take only a few tenths of a second in total. Neither of your friends notice.

The telephone, although Japanese, was made in America. The satellite is owned by an American company, which is owned by its stockholders— which may include you. It was launched by a company jointly held by a consortium of European countries. That company has thousands of employees, several of whom are Americans who live and work in America. The satellite company, nominally American, also employs thousands of people, many—but not all—of them Americans residing in the United States. Others are in Singapore and other places and may be Singaporean, Indian, British, French, or any of a number of other nationalities. Large corporations no longer owe allegiance to any one nation.

Similarly, the computers in the telephones, radio transceivers, and satellites were designed by firms in Palo Alto but use parts made in Houston, Taipei, Kobe, and Brussels. They were put together in Seoul, shipped to Amsterdam, and sold all over the world.

Unconscious—and uncaring—of all of that interconnection, Jodi hangs up the telephone and pulls a portable computer out of her purse. Brushing aside her butterdish, she types something into the computer, which beeps and makes faint chugging noises. It is retrieving information. The tiny computer can hold several hundred novels' worth of information—which seems odd, considering the tiny size of Jodi's purse. Turning to you, she mumbles that she has to fax some information to your mutual friend in Curaçao, and her computer does so as she speaks. Like the telephone, it can send and receive information from around the world; but unlike a simple voice transmission, this information can take the form of handwriting, pictures, music, or just about anything else. She pays no attention.

At the close of the meal you hand your waiter a credit card. As he wands it, your credit information pulses down the restaurant's telephone lines to the credit approval center, a bunch of people and computers you've never seen. You haven't, of course, paid for your meal; you've promised to pay for it. Actually, the credit card company has promised to pay it, and you've promised to pay them. Later that month you might send a signed piece of paper, make a telephone call, or visit a money machine—to talk your bank's computers into paying your credit card company.

Your bankers are another group of strangers, and most of them are computers too. They pay the credit card bill—or, rather, promise to pay it—because the government has promised to replace any money lost up to a certain amount. The government—that is, everyone in the country—has promised to pay the bank because national banking insurance is part of the social contract worked out to make this vast web of dependencies function.

This web is but part of a larger web that includes every nation. While sitting at the dinner table, you—or rather some of your possessions—are affecting decisions being made around the world. For instance, the money in your bank that nominally belongs to you isn't actually in the bank's vaults. It is joyriding the airwaves. Even when you're asleep, your bankers—or more usually their computers—are trading on your money, using it to generate income, a tiny bit of which they'll later give to you.

People and machines you've never heard of in Hong Kong, Tokyo, London, or Chicago are continually buying and selling things, effectively, with your money. By their separate actions, they are collectively determining what future projects will get funded: a corporation's supertanker, a country's space program, or that backyard swimming pool you've been thinking of getting a loan for.

They all depend on your money, although, like information, it is only really profitable when combined with the money of millions of others. Now that it's gone electronic, money—like information and labor, and all the rest of the strands that join together the global village—is a fiction built on top of a fantasy masquerading as a reality. As insubstantial as it is, it, and all the other strands of the web, binds us all. We are all caught in a huge yet invisible web of connections, each of us made more powerful by the uniformity and collectivism of shared life. Food, fuel, jobs, money, artifacts, life-styles, information—all are shared. Computer networks are but the latest strand in that web. Eventually though they may become the most binding thread of all.

The World's Largest Conversation

Millions of people are now sharing information on the world's computer networks. Currently an estimated eighty million people in ninety countries have computer network accounts. Perhaps ten million of them use the net regularly. Things are changing fast right now; so, instead of picturing the detailed steps of interacting on the net today, imagine that it is a giant building where you can instantly teleport yourself from room to room on a whim. There is, as yet, no notion of its distances, location, or scale.

In this imaginary building the size of London, every two of the people in the building could talk to each other if they wished. The city-sized building is divided into wings, the wings into floors, and each floor into millions of rooms. Some of the rooms have a big blackboard that anyone can scribble anything on. The posted information can be a written message, an image, a movie, a whale song, a conversation—anything. Anyone looking into such a conference room can see everything on its blackboard.

In other rooms of the giant communication building, groups of people you can't see—and who can't see you (at least today)—are chatting. They're all madly scribbling on the same blackboard at the same time, each one in a separate little area. They can all see what everyone else is scribbling, so it's a kind of ongoing written conversation—a visible chat.

In thousands of other rooms, people are playing very specialized games, writing a kind of continuing, do-it-yourself novel. Each person is a character, or more than one character, in a fantasy world conjured into existence by the words of all the other participants. They're having an experience—a very addictive experience.

Still other rooms are sites of town meetings. The residents of these towns may have never met physically and may live all over the world. But they are all passionately interested in a certain topic—perhaps gun control, tea drinking, computer crime, recycling, or white supremacy. The list of topics goes on and on. From bondage to beauty queens, from Tibetan ashrams to drug abuse, from action films to Japanese haikus—whatever your kink, on the net you'll find thousands, perhaps millions of others who share it.

Some of those following a discussion never contribute anything to it, and many net denizens follow numerous discussions. People can create new discussion rooms if they want to, once they find others to talk to. If they wish they can make their conferences private, so that outsiders have to be initiated to be let in.

Everyone on the net also has a private lockable room. Inside it you can have complete privacy if you wish, although it has a pneumatic tube others can use to send you messages. Those messages might be anything—a brief note, a video conference call, a movie extract. You can choose to read the messages, destroy them, edit them, save them, repost them, or do whatever else you wish with them. You can send mail to yourself, to a specific person or group, or to one or more of the blackboards in the building. On the door of your room you can also post information you want browsers to read. Such messages can be of any length—even whole encyclopedias' worth—and they can be so interesting that hundreds of thousands of people, instead of joining the various conversations going on around the building, simply roam the corridors, teleporting themselves from door to door and reading the information posted there.

Thousands of rooms aren't owned by people at all but are used by governments and corporations. AT&T is there, as are the White House, Fuji Electric, many high schools, museums, and libraries—even some vending machines. Their doors, like yours, display information they want the world to read. In all, there are millions of rooms in the city-sized building and, like everything else about the net, their number is doubling every few months.

The net is already bigger than the population of Canada and Australia combined. At its present growth rate, when it doubles it will add the populations of Tokyo-Yokohama (about twenty-eight million people), Mexico City (about twenty-three million), and Sao Paolo (about twenty-one million). And then it should double again. The net is a coffeehouse the size of Wyoming.

And as in a coffeehouse, no one is in charge on the net. Rank, appearance, and social standing in the outside world mean nothing there unless they help guarantee someone's knowledge. Denizens of this vast building work at their own pace, choosing the size and timing of their postings and replies or simply sitting back and reading. Those who only read are there for amusement, or information, or bonding. Those who post messages are there for the same things, or for self-display, prestige, friendship, or an argument. Everyone is there for communion.

This is a social gathering different in style, scope, and scale from anything we know. It isn't a cocktail party, although it's a little like one. It isn't an office meeting, although it can sometimes seem like one. It isn't a picnic, a party, a riot, a face-to-face conversation, or a get-together around the water cooler. Nor is it a bunch of notices thumbtacked to a bulletin board or graffiti scrawled on a toilet wall. It's all of the above and none of the above. It's a new thing—a groupmind.

The net is a place to make friends, annoy or amuse others, sell stuff, gain attention, lose respect, display knowledge, learn new things, and think about reality and your place in it. It's an electronic hivemind, a breathing encyclopedia, a living Talmud, a source of social comfort, a flea market, a place of ideas, a place of amusement, and a place for communion. It's the net in the nineties.

Drinking from a Firehose

Computers can let us create and interact with a vast sea of electronic information. In the future, every person, business, and government in all advanced countries may have available vast stores of electronic information—some public, some private. By then information will have become the blood and sinews of society.

Even now, the information ocean everyone swims in is so vast that no one can hope to find anything without markers and directions. Eventually we might have to assign all information nodes fixed locations, perhaps on a vast grid. Each node might then appear as a unique object everyone can recognize while ambling around the electronic landscape. For some, navigating through the information ocean might be like driving a car and using landmarks for directions. Others won't bother to do the navigating themselves but will have their electronic helpers do it.

Eventually, each node might be created by hundreds, thousands, perhaps millions, of computers, all working together to maintain a particular information base and to provide the fixed visual form of it that passersby can easily identify—like the store signs and neon billboards of today.

A primitive version of that electronic information ocean is forming today. Like liquid mercury poured onto a plate, present-day computers are rapidly pooling into vast communication networks spanning the globe. The people using those coalescing computers all talk to each other, producing a blizzard of electronic mail messages every day—an estimated six thousand million in 1992 alone.

In 1994 roughly a thousand million people around the world, almost a sixth of the world's population, watched the American Superbowl. Another thousand million watched the Academy Awards that year. People from Japan to Mexico to Holland regularly watch American movies in English. We're becoming the global village Marshall McLuhan predicted several decades ago.

Yet, while many of us watch television broadcasts today, few of us can produce those broadcasts. What might it be like when a thousand million of us can communicate with each other directly—without the need of filters like radio shows, newspapers, magazines, or television? If computer prices keep falling yearly (very likely), and if computer networks keep doubling yearly (very unlikely), we might be able to do this in under fifteen years. Hundreds of millions of us in advanced countries might carry around cheap, robust computers that can talk directly to the worldwide network of computers. Millions of people already carry pagers; they could easily become portable electronic mail creators instead of simple message accepters. All we have to do is add a small keyboard, or wait for cheap computers to improve enough to understand more of our speech and handwriting.

Soon, if we choose to, any one of us could tell the entire world what we think about anything, anytime, anywhere—not simply by voice transmission (as in most of today's telephones) but also as home movies, call-in talk shows, and personalized newspapers. There are already a few video cameras linked to the net providing a daily feed of various remote locations—a coffee machine, an aquarium, a park, an office. Thousands of people watch these nearly unchanging scenes from around the world; these spectators are kin to those who squatted in front of early television sets, endlessly waiting for the first broadcast. Nothing stops anyone from putting on a show—or reporting the news—in front of one of these cameras. Eventually, they could even charge admission. What might it be like to live in a world with the equivalent of a hundred million unregulated television stations?

Who needs CNN when fifty ordinary people at a riot can uplink what they see as it happens? How might that affect repressive governments wishing to control their populations? What might it do to little old ladies walking the night streets armed with satellite cameras?

Making everything electronic—numbers, words, pictures, sounds, maybe one day even smells and tastes—is like inventing currency exchange rates. Nowadays, sound is rapidly turned into electronic pulses. Soon sight, then perhaps pressure, heat, and even taste and smell will follow. All are merely different ways to transmit information and can be coded electronically. All can be exchanged. Free currency exchange led to the world economy; what will free electronic exchange lead to?

Once something can be exchanged easily, large masses of it can collect in various places. And large masses of anything—a river of water, a barrel of money, an army of people, an ocean of information—can be used to do previously impossible things. A country's space program, for example, simply couldn't exist without a liquid monetary system that lets us pool vast resources. What might today's pooling of information make possible tomorrow?

The potential social changes in ideas of community and identity, allegiance and cohesion, are—literally—incredible. What will be the effects of tomorrow's communication technology on today's entertainment industry? Or global politics, for that matter? Could instabilities in one country eventually come to be mirrored instantly in another? Are we headed for a world of instant global rioting? What would it be like to throw a party and have half a million people come?

Gossip Makes the World Go 'Round

A few hundred thousand people now carry portable computers and use them to link up with worldwide computer networks. What these people do today suggests interesting social and political changes for all of us in the future. For example, armed with today's computer technology, many corporations judge our creditworthiness, insurance-worthiness, and so on. So why don't we, in turn, judge them?

Today, most of us identify good and bad restaurants by asking friends or by reading restaurant reviews written by experts. But taste is a variable thing, and experts are both expensive and scarce. What if everyone who ate at a restaurant could post a quick review to the electronic networks for anyone to read? Bad businesses might die faster, and good ones might do better faster. Electronic word-of-mouth could let everyone consult past customers' opinions.

If that sort of thing could work for restaurants, why not shops and banks? For example, some groups on the computer network currently post reviews of travel agents, illustrating them with accounts of personal experiences. But if travel agents, why not doctors, lawyers, and stockbrokers? Doctors have enormous power over patients, partly because of their extensive training but also because of their continuing sources of information. They have far more information about us—both in the mass and individually—than we do about them. The same is true for priests, and lawyers, and police officers. That's why we have laws preventing these privileged groups from exploiting their special advantages.

What happens, then, to that ancient balance of power when anyone can connect to a worldwide, or at least citywide computer network and read other patients' opinions of various doctors and treatments? Or of bankers, claims adjusters, real estate agents, stockbrokers, and government officials. What kind of world might it be where any one of us can have almost as much knowledge as doctors and banks? Electronic networks of communicating strangers gossiping about who's good and who's not, what's good and what's not, might empower a true democracy—a true global village. But it will surely also destabilize the ancient power relationships we today take for granted.

In any case, such a paradise probably couldn't last. Inevitably, some contributors will become more respected than others, either through superior communication skills, greater trustworthiness, or higher social standing. They will then implicitly set standards of acceptability for various things—just as *Variety* sets standards for many movies and the *New York Times Book Review* sets standards for many books. Some of the most respected critics will then arrange to be rewarded for their efforts. Most of us, relieved from having to sift the good from the bad, will pay. Of course, some reviewers will sell out or be bullied out, but a few will stick to their intellectual guns. Perhaps they'll publish their electronic views in something called *The Hourly Rant.*

Further, we could evaluate people just as easily as businesses and professionals. If you met some people you wanted to know, you—or more likely your electronic pets—might dig up everything they had written on the computer network and everything anyone else ever said about them professionally or informally. Think how social relationships might change if prospective partners meeting at a bar could check up on each other while still at the bar. What might become of our notions of personal privacy?

On the other hand, not everyone will have the money, technical competence, or background to buy and use the new machines. Even today, some people in advanced countries don't have telephones. But a telephone is one thing; a line into near-priceless information sources is another. What will become of these people?

Will the technologically adventurous continue to accelerate the difference between the increasingly enormous amounts of information at their fingertips and the relatively meager amounts at everyone else's? If so, what might become of our societies? What about the huge populations in poor countries who can't even afford telephones? Will we correct the massive differences between what they can do and what we can do? Or will the disparity simply grow worse?

The Politics of Power

In 1989, a fifty-year-old empire began to fall apart. Two years later it was dead. A wave of change crashed over Eastern Europe, crumbling old power structures, cracking the Berlin Wall, and leaving power vacuums swirling in its wake. Poland, Hungary, Czechoslovakia, and East Germany changed their governments in weeks. Even the brutal autocracies in Romania and Bulgaria collapsed almost overnight. The Soviet Union was brought to its knees in part because its command economy couldn't keep up with the enormous cost of modern war machines. But new communication technology played a part as well.

Governments know that arming their populations is only wise if they won't turn their guns on you. For example, in the old Soviet Union it was a crime for an individual to own a copy machine—because it could be used to incite a revolt. Today fast, cheap, reliable communication technology is beginning to erase ancient political boundaries in a way that the gun and the tank never could.

What was true for copiers was also true for personal computers, printers, video cameras, videocassette recorders, home satellite dishes, fax machines, cellular telephones, and all the other techno-paraphernalia of computation and communication today. All were made possible by computer technology, and all were severely restricted in eastern Europe. Restricting such devices, however, created profound disadvantages. Those economies then lagged even further behind those of more open societies, until one day there was simply not enough of everything to go around and the people revolted.

Of course, that's an oversimplification of a complex process. Democratic countries probably won the Cold War because their economies were robust enough to win the last ruinous arms race. Nonetheless, it's no coincidence that personal computers and other choice-enhancing, information-distributing devices were becoming widespread in advanced countries by 1981 and that ten years later European communism was essentially dead.

Computers are not just toys. People who learn to use them (or at least, not to fear them) can turn their knowledge to productive use. This ability is useful on the factory floor, where robotics, more-efficient planning, greater flexibility, and enhanced predictability soon become the norms. The Soviet Union, by keeping such technology from its population, was trying to run a race by first breaking its legs.

Computers don't just benefit industrial production; they also make possible unprecedented improvements in communications. And that always has political consequences. During the 1991 attempted coup in Russia, for example, programmers used their computers to keep in touch with the rest of the world, even though the insurgents controlled the centralized radio, television, and newspaper facilities. Messages traveled from Moscow to Vladivostok, to Berkeley, to London, and back, while the technologically illiterate old-timers were powerless to stop them. In the old days, it was easy for Moscow to prescribe what the entire country thought simply by controlling the central broadcasting stations. Not anymore. This time, millions watched the coup unfold, and the glare of publicity proved too bright.

The wave of change that began in 1989 was also encouraged by the inexorable creep of these same technologies into inflexible regimes, despite governmental discouragement. Through gifts and smuggling, eastern Europeans with satellite dishes could finally see and hear for themselves what goods were freely available in the West. And they wanted them too. The Berlin Wall could keep the people in, but it couldn't keep the satellite broadcasts out. After living in what was essentially a wartime economy for fifty years, and finally seeing Western television, eastern Europeans decided that they wanted Coca-Cola, and they wanted blue jeans, and they wanted freedom. So perhaps the end was in sight for the Soviet Union as soon as the first cheap videocassette recorder rolled off the assembly line in Japan.

These same technologies also have consequences for the West. Our current structure of government, for example, is a holdover from the days when it was a day's or a month's journey to the nation's capital and most people couldn't leave the farm to vote. We therefore chose representatives to serve as our voices on the national scene. These representatives then chose someone to represent the overall body, and so on up the chain of command.

That hierarchy is no longer essential. Geography is irrelevant in the computer age. It might be possible, for example, to have all major decisions voted on by nationwide referenda rather than electing representatives and having them vote. Such a system sounds more democratic. But what would it do to politics?

Well, for one thing, power brokers would be very annoyed. Instead of influencing a few people they would have to influence millions. Of course, the politically cynical among us will conclude that power brokers will simply spend more time and money influencing the vote through the mass media. Besides, once a vote is taken, it will still have to be implemented, and we can't be involved every step of the way—we bore too easily. Politicians, it seems, will have an interesting life for decades to come.

Even now, politicians around the world know they usually don't have to appeal directly to the voters; they only have to appeal to the business and professional elites who really run the country. The more money you have behind you, it appears, the higher are your chances of winning. So it may no longer be one person one vote, but one dollar one vote. Perhaps we no longer hold elections—we hold auctions.

Grave New World

Strange things are in store in the future. Today's portable telephones fit in a shirt pocket, and we can use them anywhere in the city. Eventually, some may be implanted in our bodies and we'll be able to use them anywhere in the world. You could call Mom from the jungles of the Amazon or the Gobi Desert, and she could answer you even if she's shopping in Tibet or white-water rafting in Colorado. You could reach anyone, anywhere, anytime. Future communication might as well be telepathy, since people with implanted telephones could carry on conversations invisible to everyone else.

As we hurtle into the future, technology may make possible changes so drastic they'll have no visible connection with what came before. Imagine, for example, a world in which suing a doctor means suing the diagnostic program the doctor used. Imagine a world of greater financial instability and even shorter boom-and-bust cycles as government regulatory agencies designed for a slower, more leisured era utterly fail to keep up with the speed of international electronic money transfers. (Even as you read this, all the money you own is chasing everyone else's money around the world, twenty-four hours a day.)

Imagine a world where you can instantly alert the police if you are threatened with assault, sending them your exact location and a video of the potential attacker. Not even masks or darkness would help the mugger if your personal computer has infrared cameras and image-reconstruction software.

On the other hand, imagine real time cosmetic software that interposes itself between you and the camera of your picture phone, so that callers see only the face you want them to see. Imagine a world in which no news is trustworthy since any sound, image, scene, movie —including those with apparently live-action famous personages—can be complete fiction. Imagine too a world of little or no privacy, of even greater earning power for the computer literate, of even bigger disparities between the haves and the have-nots, of wholesale social disruption as new and ever more volatile technology boils through society.

When mail is delivered in four-tenths of a second instead of four days, would you like to be a postal worker? What will it be like when the proportion of the work force in manufacturing in developed countries— now roughly 20 percent—drops to 10 percent—less than four times the 3 percent presently employed in agriculture, which itself once occupied over 90 percent of the population? In such a world postal workers may have lots of company.

Everywhere Is Here

In 1990, Florida police investigating several vampire-style killings found a map leading to a body—in a suspect's computer. Although authorities never found the body, the suspect pleaded guilty to rape. Police say he lured victims to his home through a computer network. In another case, a California man operated a computer bulletin board to lure young victims to his home. One policeman logged on posing as a fifteen-year-old boy, while another signed on from Pennsylvania pretending to be a fellow pedophile. The suspect was simultaneously corresponding with the California "boy" while sending the Pennsylvania "pedophile" explicit messages about his intentions. Police nabbed him when he tried to attack a police cadet sent to his house undercover.

As the net's demographics broaden, more fraud artists, rapists, and killers will inevitably enter it. The crimes they will then commit will be as much computer crimes as bank robberies are gun crimes. The more we use computers, the more we will find ways to abuse them. We're rapidly approaching the point where we can live in both a physical world and an electronic world. And in both, we will be doing exactly the same things.

Today's net is a frontier boomtown, with all the energy and all the vice that entails. It breaks down all traditional jurisdictions and changes all the rules. Walking a new beat in a neighborhood without an end, today's police, used to nabbing crooks with fingerprinting and brute force, are now learning to track electronic footprints across continents. We will, inevitably, try to map crimes on the net to our present understanding of the law. But the map won't be very accurate. Until now, everything happened somewhere. Now even the most basic things, such as where a crime occurred, is open to question. Is it where the assailant is? Where the victim is? Or somewhere else? Where should it be tried? Whose laws should be used?

Questions that once had clearcut answers are now blurring into meaninglessness. Who should be involved in a computer chase? Who has jurisdiction? If you invade someone's computer, is that burglary or trespass? Where should the search warrant be issued? And what for? What happens if someone living in country A commits a crime in countries B and C using computers in countries D, E, and F? What are the rules?

The Empire Strikes Back

Consumers are the consumed in a consumer society. Corporations and governments, through advertisers and the mass media, buy chunks of consumer attention the way we buy bars of soap. Mass communication providers, essentially, sell us to ourselves.

But for most net denizens—at least today—the net is something quite different. People use it to reach out to others and ask or answer the same questions we all implicitly ask: What's it like there? How do you live? What makes you laugh? Why do you cry?

Censorship is a wedge corporations and governments often use to split the blocks of social solidarity resulting from such exchanges. All through the recent past, whenever a new and unregulated communications channel has appeared—a broadsheet, the telegraph, television, or a computer network—business and government have seen it only as a tool of corporate revenue or social control.

Some of the new channel's users don't want that control. They want the freedom to say anything they please, to anyone they please—which, inevitably, displeases others. The unexpurgated consequences of normal human life—sexual, religious, racial, and political—frequently offend them and they complain. The power brokers use those complaints to quash the freedom of the new medium. Eventually, it becomes as controlled as all the old media. That's the way it's always been. Despite the present tidal wave of freedom, it will probably happen with computer networks too.

Television was once seen as a similar brave new frontier of widespread education and justice and equality for all. Soon, though, it turned into a friendly monster that sidled up to us every evening and, by dangling pretty images to distract us from the earnestness of life, slowly, methodically, began to eat our minds.

Still, there's some hope for those who believe in freedom of expression. Unlike all past media, a computer and a telephone line are the only things you need to participate in the discussion. You don't need a printing press or a television station. Eventually what you need should become as small, as cheap, and as common as a telephone. Although most governments still regulate the telephone lines, no one, at least publicly, monitors the content sent along them. And with a telephone you can only speak to one, but with a computer you can speak to millions.

Further, the net, as it exists today, is extremely decentralized. If two people want to communicate and someone blocks a line between them, there's always another path to connect them. Today's net handles censorship the same way it treats a broken connection—by routing around it. Of course, if some corporation or government manages to seize control of the net, that could change. As yet, however, it isn't tied to big commercial interests. It's still relatively cheap—at least for the educated middle classes of advanced nations. So its content remains largely unregulated. It isn't controlled from above, because there's no "above" to do the controlling.

On the net today, you can say anything you wish to anyone you wish. Theoretically, in a free society, that shouldn't make any difference; you supposedly always have that right. But it's one thing to get up on a soapbox in the park and speak your mind to a few dozen people and quite another to express your opinion to a few million people.

Before the net, only a few people, backed by governments or big corporations, could do that—and they had to spout the party line. In an age of mass broadcast media alone, a few large vested interests could control what was said. They told us what to think—or at least, what to think about. That's not so easy when there are millions of cheap, decentralized communication channels. On the net, theoretically at least, millions can hear you—no matter who you are and no matter how little social or economic power you hold. And that, perhaps, is the foundation of freedom: the freedom to speak your mind.

Pregnant with Possibility

Only connect! That was the whole of her sermon. Only connect the prose and the passion, and both will be exalted, and human love will be seen at its highest.
E. M. Forster, *Howards End*

Today's computer technology is rapidly turning us into three completely new races: the superpoor, the rich, and the superrich. The superpoor are perhaps eight thousand in every ten thousand of us. The rich—me and you—make up most of the remaining two thousand, while the superrich are perhaps the last two of every ten thousand. Roughly speaking, the decisions of two superrich people control what almost two thousand of us do, and our decisions, in turn, control what the remaining eight thousand do. These groups are really like races since the group you're born into often determines which group your children will be born into.

Only the rich and superrich have the opportunity, education, and resources to own and use computer technology. It's hard to get excited about computers when you're busy starving to death; or living in a tense matrix of drugs and crime and fear; or being forced to prostitute yourself for food and shelter; or being repressed by your own government.

A boy in Bangladesh, a girl in Somalia, or—perhaps saddest of all— many ghetto children in the world's richest countries, might never be able to step onto the world stage that is the net. Working their way among the destroyed buildings and equally destroyed social structures of their environment, they are permanently disenfranchised from the world many of us take for granted. Computers alone cannot erode that mountain of destitution.

Today, we number nearly six thousand million people, but we have only five hundred million telephones—one for each twelve of us. Half of us have never placed a single telephone call in our lives. There are already well over a hundred million computers, and over sixteen million more are added each year—with the number rising rapidly. In a decade or two, five hundred million of us may be connected to a computer network. That connection might eventually make the difference between being an information-starved peon and a global villager.

It is often said that computers will destroy ancient power centers and enfranchise greater masses of people. It is true that in America over a third of all households already have computers. Yet that proportion jumps to well over half for homes with university graduates and to three-quarters for homes with incomes above seventy-five thousand dollars. If that pattern is repeated in the world as a whole, computers might merely ensure that the ancient pyramid of power extends into a thin sharp spike, with a tiny few determining the fate of the millions.

By the time that happens, however, there should be a few hundred million people active on the net. (Just think, a global electronic community of a hundred million gossipers.) Perhaps among them will be enough people both connected to the rest of us and compassionate enough to speak for the dispossessed. Perhaps.

Every new communication medium is special. But the net is extra special, because we, the users, are the ones making it, not some faceless corporation. It's growing as fast as it is because it merges two hugely endowed technologies: computers and communications. But it's also growing rapidly because the more people who get on it, the more valuable access to it becomes. Like telephones and fax machines, tennis clubs and bars, the net becomes more attractive as more of us join it. So it's likely that everyone who can afford to will be on the net in fairly short order.

We're gregarious, gossipy creatures, continually talking to each other to find out what others think—or what we should think. Out of that reliance we build community, whether in a pub, at the laundromat, or on the net. Without that communication, that bonding, human society would cease to be. But the net presents numerous dangers too. The search for communion could easily turn into a flight from community. Don't like your neighbors? No problem; turn off, tune in, and drop into a whole other world of people you never have to actually touch. Further, as its demographics broaden and it goes commercial, the net's content will inevitably become more splintered, more trivialized, and perhaps more acrimonious. That's the way of things.

Computers won't bring about a better world—perhaps nothing can do that. But they certainly can change the world: in some ways for the better; in others, for the worse. That's the nature of today's new technology. By changing things in fundamental ways right before our eyes, it lets us see more clearly who we really are by showing us what we truly value. Sometimes, perhaps far more often than necessary, what we see is an ugly side of human nature.

Still, the net today is a glorious experiment. Considering that no one planned it and no one controls it, it shows us that we can make very complex systems work despite our many flaws. As it quickly commercializes, and changes drastically, we should work to preserve some of its more democratic and humane characteristics. Because that's something to be proud of.

Only connect.

5

The Bloody Crystal

I feel the need, the need for speed.
Top Gun [film]

The man who laughs has not yet been told the terrible news.
Bertolt Brecht

I fully expected that by the time I was 21 . . . [s]cientific truth was going to make us so happy and comfortable. . . . What actually happened when I was 21 was that we dropped scientific truth on Hiroshima. We killed everyone there.
Kurt Vonnegut, *Breakfast of Champions*

It's about five meters long, weighs about eight hundred kilograms, and costs about a quarter million dollars. It's just been launched from a French-made Argentine warplane, and as soon as the pilot releases it he speeds for home, his mission over.

It isn't a bomb. It's a French-made flying robot. It carries 160 kilograms of high explosives and the computer smarts to deliver it in the teeth of sophisticated defenses. As long as its fuel lasts, it will hunt with ruthless precision, then blow itself and its target up. Three minutes from now it will destroy a British warship.

It's Tuesday, May 4, 1982, and Britain and Argentina are fighting the Falklands War. The missile is an Exocet. Four seconds after launch it is cruising only a few meters above the choppy South Atlantic. Named for a flying fish, the ship killer hugs the sea, lost in the confusing clutter of radar echoes from the cresting waves just below. It is too low for radar detection. Its computer keeps the missile low and on course, adjusting its stubby fins in response to signals from its internal sensors and gyroscopes. Moving at about eighteen kilometers a minute, it's less than three minutes away from the British destroyer H.M.S. *Sheffield*.

At about thirty seconds from impact it switches to an active homing radar, locking on. It rises slightly to scan the horizon and check the target, making sure it's still on course and prepared to counter any evasions in hundred-thousandths of a second. But there's no need for a course change. The target is unaware of its approach. The flying shark subsides back to wave level, lapping up the few remaining kilometers at supersonic speed.

The fifty-million-dollar *Sheffield* was protected by one of the most advanced defense systems ever built. Yet it was defenseless. When the missile struck, the men on the *Sheffield* had no idea what hit them; they had had only fifteen seconds warning. "On the upper deck you could feel the heat through your feet with your shoes on," the captain said later. "The superstructure was steaming, and the paint on the sides was coming off. The hull was glowing red and hot. We had no hope of retaining the fighting capability of the ship." The destroyer was lost, and with it twenty-one of its crew.

Of the ten ships and over a hundred aircraft lost by both sides in the Falklands War, robots destroyed more than half. Smart weapons like the Exocet have a better than nine-in-ten kill ratio against unwary foes. And they're cheap. War is quickly becoming a game only machines can play.

Smooth-Running Gun

When most of us think of robots, it's usually as slow, dumb, unwieldy beasts slaving in factories, much as oxen slaved in fields generations ago. But robots aren't only in the factory, helping us make our cars; they're also on the battlefield, helping us make our wars. And they certainly aren't slow, dumb, and unwieldy.

During the 1991 Gulf War, America released 288 cruise missiles against Iraq; 80 percent of them hit their targets, with no loss of human pilots—at least, no American ones. Each missile delivered up to 450 kilograms of high explosives. They could just as easily have carried nuclear warheads.

A cruise missile is essentially a robotic bomber. It's like a piloted bomber, but better: It's smaller, cheaper, stealthier, and very precise. Its computers record exactly where it is at all times, thanks to radio contact with world-spanning satellites; and it has extensive electronic maps of where it's going. Like a bloodhound straining at the leash, it can be launched more than two thousand kilometers from its target and reach it to within a few meters. It could hit a garage door in Dallas from a rooftop in Los Angeles.

Flying at treetop level, it can turn corners, avoid obstacles, and fly through doorways. Its small size and low flight path lets it hide from enemy radars in the radar clutter near the ground. Even if detected, it's almost impossible to hit. Its next generation, to be deployed during this decade, will be smarter, stealthier, and faster and will peel off independently targeted smart missiles and bomblets as it flies.

Since each cruise missile has its own computer—and satellite links to establish its exact location—a cruise missile salvo can regroup in flight like a flight of deadly birds to hit its targets in precise sequences. The first wave usually destroys the foe's radar and surface-to-air missiles. Later waves mow down the installation itself—a missile complex, an air base, a command bunker. Next-generation missiles will adapt their targeting in flight as conditions on the battlefield change and as targets move around.

Today's cruise missiles cost about 1.3 million dollars each. Since much of that cost pays for sophisticated electronics, the price will drop as computer technology matures. A single American stealth bomber, on the other hand, costs hundreds of millions. And that's not counting the costs of employing several dozen mechanics per plane, training flight crews, and transporting many tons of fuel, munitions, and support equipment. It's certainly not counting the human and political cost of losing even a single pilot. Although nowhere near as adaptive, robot planes are cheaper and decidedly more expendable. It's clear where the future of air war lies.

Nintendo War

Gone are the days when a captain could stand on a ship's bridge and see everything important through a pair of binoculars. Nowadays, a warship's radar extends at least a hundred kilometers in every direction—from four to ten times farther than simple binoculars. Nor does its vision go away at night or in fog. Nowadays, the captain's place isn't on the bridge, it's in the warship's computer nerve center.

A darkened room of glowing consoles and tiny blinking lights, the control room may as well be underground, far from the ship it commands. During war, the only physical feedback officers there get is the noise of departing missiles and the swaying of the room as the ship heaves through the swells.

For these officers, war has become a video game. Enemies are no longer ferocious warriors with teeth bared and weapons raised. They are blips on a phosphorescent screen with little symbols showing what a computer thinks they are. If the program controlling that computer is wrong, the information is wrong and the officers make mistakes. More often though, the emotionless computer is right but its displays are badly designed; and in the fever of war the officers are likely to misread the computer's messages.

In 1988, the American cruiser U.S.S. *Vincennes* carried the most sophisticated ship defense on the planet—six hundred million dollars' worth of electronics to detect, identify, and target scores of planes, ships, and submarines within a three-hundred-kilometer radius. Unaided, it could assign dozens of missiles, torpedoes, rockets, and guns to its targets and automatically track them while waiting for firing orders. In case of all-out attack, the captain could give the system complete control of all weapons and sensors.

On Sunday, July 3, 1988, at a tense moment in the Persian Gulf during the Iran-Iraq war, the *Vincennes* was attacked by Iranian gunboats. The ship's officers, who had been warned just that morning that Iran had deployed fighter planes that might attack the flotilla, mistook a distant Iranian commercial jetliner for an attack craft.

Just a year before, an Iraqi warplane had accidentally fired two Exocet missiles at another American cruiser, killing thirty-seven and wounding twenty-one. Faced with a possibly similar threat, the *Vincennes* waited as long as possible, then shot down the plane while it was still fifteen kilometers away. Two hundred and ninety civilians died.

Four decades earlier the cruiser couldn't have shot down the airliner. Before doing so its aircraft carrier group would have had to loft a warplane to make visual contact just to aim the missile properly. The pilot might then have seen that the target was an airliner and not an attack plane as reported.

That can no longer happen. Nowadays, computers can handle all the aiming and firing. Once upon a time, ships had to board each other with cutlasses bared to decide a sea battle. Today, naval battles can be fought and won in five minutes by combatants a hundred kilometers apart. Smart weapons and long-range scanners have so distanced mariners from their battles that a warship's commander sometimes has to decide in seconds—based only on a blip on a screen—what action to take. The rule in war is that if the foe is in range, then so are you.

With the time to decide life-and-death matters sometimes reduced to seconds, we have less control. As computerized antiship missiles like the Exocet got better, ships started carrying computerized guns that can fire automatically on anything approaching the ship that looks like a missile. There's no longer time for John Wayne to smoke a meditative pipe as he plots an attack on the enemy's ships. Too many things happen too fast and too far away for warriors to make eyeball contact. As a result, mistakes happen.

On the other hand, mistakes always happen—particularly in the chaos of war. And perhaps human-computer systems make fewer mistakes today than unaided people did in the past. Unfortunately, because our systems are far more complex, powerful, and integrated, the consequences of today's mistakes can be far graver than yesterday's. When village waterwheels were our main source of power and bows and arrows our main weapons, a mistake in the way we use them couldn't affect that many of us. Things are different in an era of nuclear power stations and nuclear weapons. Blundering while hurling a spear is one thing; blundering while launching a space shuttle is another.

Cold Warriors

During the Gulf War, the United States Navy lost two major ships—a large helicopter carrier and a cruiser—because it wasn't prepared for new-technology mines. In the past, mines were simply bombs studded with contact switches and they floated just below the surface. Any ship or submarine hitting one of the switches activated the mine. The Navy had prepared for that century-old mine technology with a new helicopter-based minesweeper that dragged a sled just below the surface to detect and catch all such mines. The problem was that, besides the old style dumb mines, Iraq used much cleverer Russian and Italian ones.

Instead of floating near the surface, these mines often live on the sea bottom; and instead of simple contact switches, they have sophisticated sensors, a computer, and memory. Some of them detect passing ships through changes in sound, heat, light, magnetic fields, or water pressure and then blow up under them.

Such smart mines can play endless tricks. For example, they can be programmed to be active only at certain times (thereby often eluding minesweepers). Or they can lull suspicion by letting a certain number of ships go by before blowing up under the next one. Some blow up only when a specified number of ships are passing over them or only when a particular kind of ship goes by. Some can even move about on the seafloor, while others wait for a ship or submarine to loiter past before firing a smart torpedo at it.

To combat the new technology, advanced minesweepers take successive sound-wave pictures of the seafloor, comparing the maps produced and looking for differences. They also use tiny remote control submarines to get a close-up look and leave an explosive near anything suspicious.

But technology never stands still—particularly in warfare. If current antimine technology gets good enough, the next generation of mines will simply get smarter. Rather than blowing up any and all ships and submarines, they may recognize friendly ones and let them pass. Rather than metallic (and easy to detect) skins, they may have plastic skins, as some land mines already do. And rather than just hanging around waiting to be found, they may, one day in the far future, hide inside moving schools of fish.

The conclusion is plain. Putting a computer in a weapon makes it much more formidable. Navies may eventually have to resort to more and more mobile and independent antimines until every important ship moves around within a protective cloud of antimines. Such a school of independent attack weapons might also be very useful against a ship's most deadly foe—the submarine. The sea will then be as alive with mobile smart machines as the air is now becoming.

The Whites of Their Eyes

Just before dawn on Tuesday, April 15, 1986, eighteen American fighter-bombers were en route to Libya. Flying low to avoid radar detection, they were tasked with destroying some barracks, a seaport, and an airport as reprisals for terrorist attacks. But because the targets were located in and around the heavily populated city of Tripoli—and because America and Libya weren't actually at war—the bombers couldn't simply carpet bomb everything and leave.

During the Second World War, there were no electronic computers. There was nothing small enough and smart enough to fit in a bomb and do anything useful. So fifty years ago, all effective bombing was carpet bombing, because all bombs were dumb bombs. After release, they simply fell.

During the raid on Libya, however, each bomb had its own computer and movable fins. When a bomber pointed a laser beam at a target and released its bombs, each bomb's computer homed in on the illuminated spot by adjusting its guidance fins. The bombers were moving so fast, however, that pilots only had about twenty seconds to find, pinpoint, and illuminate their targets. During that twenty seconds they could also be under heavy antiaircraft fire. So, the bombers were supposed to sneak in low and fast to avoid radar detection. Then—in twenty seconds, more or less—they had to pinpoint the targets using heat-detecting cameras, illuminate them with their lasers, release the bombs, keep the lasers aimed until the bombs hit, then skedaddle for home.

That isn't quite what happened. America chose not to use its then-secret stealth fighters and instead sent radar-detectable warplanes on the mission. The Libyan military picked them up on radar and attacked with missiles and gunfire. As the pilots tried to evade the fire, some of the laser beams drifted far off their targets, and the bombs blew up whatever the lasers illuminated last—in one case, part of the French embassy in Tripoli.

War is never like the simulations of it before the battle, nor like the sanitized versions shown on television. Opponents have a stake in winning too, and try hard to win. There would be no war if that weren't so. Computers or no, war will always be a high stakes contest of attack and counterattack, feint and counterfeint.

Between the 1986 airstrike on Libya and the 1991 Gulf War computers grew ever cheaper, faster, and smaller, and smart bombs became more independent. Now bombers need only illuminate their targets for a few seconds to let the bombs pick up the video images, analyze them, and home in on their own. Other, even more precise bombs carry a rocket in their tail and a video camera in their nose so that pilots can guide them from far away. Smart bombs are now smart missiles, and the time to aim and fire them effectively has dropped from twenty seconds to five seconds.

Thanks to computers, weapons are becoming smarter and more independent. Once upon a time, our weapons were so slow and inaccurate that we had to wait until we could see the whites of their eyes. Now we don't even have to wait until we can see their country, let alone their eyes.

The Vanishing Pilot

Over half the cost of today's advanced warplanes goes into computers. With fire-and-forget long-range smart missiles, most aerial dogfights have become contests to see who has the best ones. Today's warplanes fight from ten to forty kilometers away—sometimes even from two hundred kilometers. Most advanced jet planes are simply unflyable without computer aid.

Warplanes are getting smarter too. They have sensors built right into their skin and they carry more and more competent computers, some of which understand and accept a few spoken commands, even when the pilot is under stress. Target-finding computers can pick out potential attack points hundreds of kilometers away at night and in bad weather. Still more computers track and relay target information to yet other computers that order the plane's missiles and guns to seek and destroy the targets. Other computers manage fuel consumption, check for hostile radar, continuously manipulate the warplane's numerous control surfaces for best flying, and plan the shortest and safest flight path.

Yet the present generation of fighter planes is not as good as it can be because pilots are fragile. Carrying a pilot today is like going to war with eggs in your pocket. Turns have to be wider and dives and climbs shallower than they could be. No such limits apply to the robotic antiaircraft missiles trying to blow the plane out of the sky. Further, during a dogfight, warplanes could be closing at over Mach 4—upwards of one and a half kilometers a second. With today's air-to-air missile range of around fifty kilometers, two pilots who detect each other from sixty-five kilometers away have less than ten seconds to decide what to do. And what's an eyeblink to us can be an eternity to a computer.

There's another reason we're vanishing from warplanes: We cost too much. It takes between five and ten million dollars to train a pilot. And the costs of a human crew add another one to five million dollars for extra space and weight (oxygen, extra fuel, seats, parachutes, and so on). In addition, warplanes are bigger, slower, heavier, and clumsier because of their human cargo. Having to carry a crew decreases an aircraft's weapons load and maneuverability, expands fuel needs and cost, reduces its range and speed, and vastly increases the human and political costs of losing even one. For all these reasons, pilots and weapons officers in today's advanced warplanes are an endangered species. They're on their way out, shouldered aside by lighter, cheaper, faster, more expendable, and more rugged silicon soldiers.

In the 1940s, heavy bombers needed a crew of twelve; by the 1950s, that dropped to six; by the 1970s, it was four; and by the 1980s, only two. In the decades to come, there will be one, and then—like cruise missiles—none.

To keep pilots in warplanes, we might eventually have to encase them in body-fitting padded coffins to compensate for high-stress maneuvers. Then, to jump up their reaction time, we might have to give them drugs or a direct brain-to-plane connection, or both. We would also have to bear the enormous costs of those sophisticated planes and their ever more stressed pilots. Even then, as the decades pass and ever better and cheaper computers continually improve antiaircraft sensors and weapons, the demands of the cockpit will eventually grow beyond the limits of human adaptability. Computers will push pilots right out of the plane.

The days of the First World War flying ace armed only with a cigarette, a debonair smile, and a plucky biplane made of canvas and wire are gone forever. The ever-increasing numbers of smart missiles and drone aircraft are pushing us kicking and screaming into our future. Soon we won't be fast enough, cheap enough, or rugged enough to stay in the cockpit. Machines are taking over. They have more of the right stuff.

The Last Boys' Club

Today's technological infantry is armed with personal computers, fax machines, video cameras, global positioners, laser rangefinders, cellular phones, day-and-night vision, battlefield screens—oh, and, of course, guns. Computers make all those new devices possible. Yet, however dressed up with the latest toys, war is about killing and avoiding being killed. That ultimate reality is becoming easier for civilians to ignore as computers invade the battlefield, distancing all except the dying from what war is really about. Still, the one certain thing that all wars produce is cemeteries.

For the infantry, war means freezing cold or baking heat and relieving yourself out in the open. It means hunger, and bone-deep exhaustion, and no hot meals for days. It means living with lice, fleas, rats, and dirt and grease so thick you feel you'll never be clean again. It means worry about your family and seeing your friends butchered right in front of your eyes. It means constant muddle and low-grade pain liberally mixed with bursts of mind-numbing terror.

Throughout the centuries, the infantry has always suffered the highest casualties of all military units. Even during this century, a combat infantry soldier's chances of being injured or killed during wartime have been better than two in three. Technology may drop that percentage over the next few decades for rich nations, but it won't ever bring it to zero. War is about causing mass death. Now that death is coming to men and women alike.

New technology, invented for one quite narrow reason, almost always has unpredictable consequences. During the 1991 Gulf War, for instance, over twelve hundred American troops left the war early—because they were pregnant. It doesn't require big muscles to pull a trigger or command a combat computer. With smart standoff weapons, it's no longer even necessary to see the enemy, far less wrestle with them. And the days when troops had to march for weeks or months carrying enormous loads are fading—at least in advanced militaries. Instead, soldiers get to the war in armored vans and helicopters. So the traditional military advantages of higher male strength and stamina are decreasing, being slowly equalized away by technology.

Nonetheless, except for aviation, sailing, and artillery, women are still denied direct combat roles in Western militaries. That doesn't, of course, keep them safe. During the Gulf War, several American female soldiers were killed or injured, and two were captured. Every soldier in a combat zone, even those in administrative or support positions, goes armed and is a potential casualty. Technology helps, but it doesn't keep you from being a target. In war, everyone is a target.

Eyes in the Sky

Over a century and a half ago, the famous Prussian military analyst, Karl von Clausewitz, spoke of the "fog of war." He meant that in war, even quite ordinary bits of information—like where the foe is—can be impossible to get. The whole point of a battle is to kill your enemies, but you can't do that if you can't find them first.

To combat the fog, advanced armed forces depend on orbiting telescopes. These spyglasses can cost more than five hundred million dollars and are probably far superior to the civilian *Hubble* telescope. And, while the *Hubble* looks up, all of them look down. Circling the earth scores of kilometers overhead, on a clear day some of these monster spyeyes can spot things on the ground as small as five centimeters across—about the width of a pack of cigarettes. Some, using heat detectors and light amplifiers, can see at night.

Besides the telescopes, there are radar and radio satellites and several microwave and laser communication satellites. The radar satellites see through clouds, smoke, and dust and can discover objects buried up to three meters underground. Some of the radio satellites can eavesdrop on cellular-phone conversations around the world. Others can detect submarines deep in the ocean by measuring heat and magnetic disturbances. Several satellites help fix the positions of things on the planet's surface to within a few meters. In the close quarters of combat, troops might need to call in an airstrike or artillery barrage on an enemy unit only a few hundred meters away. In war, it's vitally important to know exactly where you are.

During wartime, advanced nations also use three-hundred-million-dollar air traffic and ground-surveillance aircraft. Each of these planes can replace the control tower of an entire major airport. They can track, identify, and disentangle over a thousand jets flying close together. Their radar can see small low-flying aircraft from three hundred kilometers away and large high-flying aircraft from six hundred kilometers away. Others can look down on a battlefield to track and identify every moving thing within five hundred square kilometers. For the first time in history, some military commanders can see, identify, and pinpoint everything on, below, and above the battlefield.

In future wars, aircraft might sprinkle tiny, cheap, and mobile noise, heat, light, radio, magnetic, and seismic detectors all over the battlefield. Commanders could then have a soldier's view, a tank's view, a pilot's view, and a satellite's view of a battle as it happens. Such technology could also give troops a nearly indestructible cellular-phone system. Orbiting satellites and other such technology are the eyes and ears of a modern military. Without them, modern armed forces are blind, deaf, mute, and lost. All are the result of the ever-increasing speed, power, and miniaturization of computers.

So, for a time, the fog of war is lifting—at least for a modern military facing a less-advanced one. But it will be back. Because all armed forces are keen to develop a means to jam, confuse, or destroy the current detection technology. Many of the secret satellites orbiting above are antisatellites—intended to destroy other satellites in time of war. So future battlefields will inevitably become as confused as they were in Clausewitz's time and the first salvos of any future battle between modern nations will have to start in space.

On the Bleeding Edge

In war, electronics can make the difference between life and death. A basic thirty-kilometer radar in your warplane might cost a quarter of a million dollars. Doubling the price can increase its accuracy and extend your sight to forty kilometers, letting you spot attackers sooner. Doubling the price again allows you to see seventy kilometers ahead, lets your radar guide your missiles, and gives you better tracking and safer low-altitude flying. Double the price yet again—to two million dollars—and you get vision out to almost two hundred kilometers and better ability to see through sophisticated electronic defenses. Similar escalations hold for the missiles you're about to fire and the warplane you're flying in.

Even the lowly bullet might eventually get its own computer. Every modern air, sea, or land warcraft needs computers—to help it sense its environment, target and deliver its weapons, move itself around, and support its crew. Electronics goes into satellites, planes, helicopters, missiles, ships, submarines, tanks, armored vans, artillery, and on, and on, and on.

All those computers cost money. America's newest attack submarine, for instance, costs 2,100 million dollars. And its newest stealth bomber costs another 2,200 million. Such enormous quantities of money give rich nations a significant military edge. The more they spend, the safer their troops are. Every commander who has to lead troops into battle and be held responsible for their lives wants the best equipment money can buy—and no one wants to tell the taxpayers that they're trading today's dollars for tomorrow's body bags.

Before the 1991 Gulf War, everyone sold weapons, equipment, expertise, and supplies to Iraq. America and Britain both provided nuclear technology, electronic warfare devices, and very large guns. Chile sold ammunition and bombs. France sold advanced aircraft, missiles, electronics, and nuclear reactors. Germany sold chemical weapons technology and electronics expertise. China sold missiles, bombs, and artillery. The Soviet Union traded three thousand tanks and five hundred aircraft. South Africa sold large-bore artillery. Brazil sold rocket launchers and armored cars. In just five years, Iraq spent over forty thousand million dollars. Everyone made a buck. At least, until the fighting started. The resulting war then cost America alone sixty thousand million dollars.

Modern war is definitely not for the poor. During the 1980s, America alone spent two million million dollars on its armed forces. Worldwide, weapons costs alone now make up about half of all military spending, and more than a third of that goes for electronics. Even though the Cold War is over, overall military spending on electronics is rapidly approaching 20 percent of the world's military budgets. All countries combined presently spend around a million million dollars a year on their armed forces. (That's more than all warring nations spent in 1944, the year of heaviest fighting during the Second World War.) Military electronics alone will thus soon consume almost two hundred thousand million dollars a year—and the amount can only increase. Nobody wants to lose the next war.

The Hyperkinetic War

Computer technology is rapidly taking over the battlefield, speeding up the pace of war. War has become wider in scope, more precise, faster, and more lethal. The enormous killing ability of today's automated weapons—and their fearlessness, self-sacrifice, and absolute lack of compassion—is, literally, inhuman. The ferocity and speed of contemporary war involving advanced nations isn't usually seen by Western television crews only because they aren't on its receiving end. The next major war will fix that.

When the Soviet Union collapsed in 1991, the East-West arms race went with it. As military budgets are slashed, more of the newer weapons will come to depend on civilian advances in computer technology, which will make advanced weapons cheaper, smaller, and more common. Because of cutbacks in military research, western forces will upgrade their weapons with the same civilian technologies the rest of the world uses; so when the next major war comes, they will face enemies armed with similar weapons. That will put a new face on modern war—and it's coming soon to a television set near you.

In the 1990s, the ratio of robots to people on the battlefield is still very low. But it's early days yet. It was only about a decade ago, for example, that the first modern cruise missiles were deployed, and they've already been through three generations—each one better than the last. Walking and crawling robots may not be common yet, but the proportion of automated systems on the battlefield is already high and getting higher every year. Guns must now be self-aiming; bombs, self-flying; and missiles, self-guiding.

The overall direction is clear: To have a chance, troops must be armed with smart sensors and weapons. They must also specialize more and be far better trained. Today's complex weapons, compared to the primitive weapons of the Second World War, are more heavily electronic and far more expensive. They take much more skill and training to maintain in the field. Further, making them in the first place requires a first-class economy. At four to six million dollars each, an advanced tank, for example, is now more expensive than some aircraft. Running a real one—like flying a real plane—can cost a lot in fuel, maintenance, and spare parts. Yet soldiers need to practice, particularly since war has become so technical. Electronics helps them do that.

Training troops using electronic simulators is invaluable to those nations rich enough to have them. They help militaries do something they previously could do only in an actual war: separate competent from incompetent officers. Many peacetime forces the world over are ill-prepared for war; their officer classes are largely made of bureaucrats and technocrats with no combat experience. In peacetime, most officers need bureaucratic ability and political astuteness. During wartime, combat officers need the ability to lead in high stress situations and a very special do-or-die mentality not encouraged in peacetime.

Further, as weapons get smarter, they're also getting more integrated and more automated. Radars in forward positions or in reconnaisance planes now relay enemy positions directly to computers in guns and rocket launchers, which automatically target them. All nations have no choice but to move to more and yet more dependence on smarter and faster weapons, smarter and faster defenses, and advanced computation and communication systems.

The common thread through all of this change is the relentless escalation of computer competence. To shoot down bothersome aircraft, one side develops antiaircraft missiles. To destroy those missiles, the other side develops antimissile missiles. To confuse the antimissiles, the first side develops jammers, scramblers, and decoys. To cut through that interference, the other side develops smarter missiles. To reduce the human cost of aircraft losses, the first side develops robot planes. And so it goes, on and on, in a never-ending upward spiral.

Evolution's Heavy Hand

Like a schoolyard fistfight, warmaking depends on six things: speed (how fast you can hit), strength (how hard you can hit), scope (how far you can hit), stamina (how long you can hit), stealth (how sneakily you can hit), and selectivity (how well you can hit). Intangibles like leadership, morale, and training are important too, but only insofar as they affect the other factors. Terrain, climate, and weather also affect wars, but they aren't (yet) under our control.

For thousands of years these six factors changed only gradually. In the present century, however, there have been three revolutions in technology and, so, three revolutions in warmaking.

The first came in the 1920s and 1930s when internal combustion engines, portable radios, and planes replaced horses and messengers and added a new dimension to war. Before then forage for the horses exceeded the amount of food and ammunition an army carried and armies fought no faster than they did during the days of the Roman Empire. After mechanization, the speed that units could advance leapt up, as did the depth to which they could penetrate while still keeping in touch with each other. Both the speed and scope of war picked up.

The second revolution occurred in the 1940s and 1950s, when long-distance communication and sensing technologies like radar and telephones and mass weapons like ballistic missiles and nuclear weapons made war strategic and up to the minute. It then became possible to quickly bring overwhelming firepower down on a distant enemy. War could be intercontinental. And it could be over in a matter of minutes.

In the Second World War, the scope of war was maximized, because armies could finally hit whole national populations. At the same time, warfare stamina was maximized by mobilizing entire populations. Strength also reached its maximum—at least until we develop planet-destroying bombs or nation-destroying biological weapons. And with the development of nuclear submarines, speed also reached its maximum—until we orbit high-energy lasers or plant satchel nuclear weapons in the enemy's cities.

The third revolution started in the 1970s and 1980s as computers grew good enough and cheap enough to start replacing us, and enhancing communications, long-distance sensors, weapons, logistics, and just about everything else about war. Computers also revolutionized design and testing, thereby adding, for example, stealth planes, missiles, and ships to the only previously known stealth craft—the submarine. The selectivity and stealth of warfare both leapt up, and weapons started, in a rudimentary way, to think for themselves. That third revolution had, and will have, many consequences.

First, cruise missiles and other smart weapons have become the new timekeepers of war. Their stealth, precision, high cost, and the long time needed to manufacture them, are making wars briefer, more expensive, more frenetic, and more ferocious. The carnage now starts with the smart weapons destroying each other, leaving the older technologies to fight to the end. Wars are quickly being decided by who can field more and smarter armaments at the beginning of the war since there is no longer enough time for any but a few nations to build new ones in time to use them. For example, Argentina's attacks on British ships soon petered out during the 1982 Falklands War because it had only five French Exocet missiles at the start of the war. Even so, those few missiles destroyed two major British ships. Iraq had a similar experience with its relatively small supply of Russian Scud missiles during the 1991 Gulf War.

Second, advanced nations, who can make and improve smart weapons easily, polevaulted ahead of other countries. On the other hand, compared to nuclear weapons, these devices are often smaller and cheaper and need less maintenance. So developing nations can buy and use them even if they can't make them.

Third, buying rather than making smart weaponry means buying a two-edged sword. Thanks to the immense complexity of the computers controlling them, it's now possible for weapons sellers to program invisible safeguards into their weapons' and sensors' software so that they can't be used against the seller's nations. Buyers might only become aware of these failsafes when they go to war and watch their expensive equipment blow up.

Fourth, as the target selectivity of smart weapons improves, the value of tactical nuclear weapons decreases, since precision weapons can now be just as effective. There's no point nuking masses of tanks or ships when conventional weapons are selective enough and cheap enough to do the job. High explosives are cheaper and safer than nuclear weapons, both militarily and politically.

Finally, the greatest change in war in the past few decades has been the speed of improvement in machine competence. Because computers now control everything in sight, a weapon's usefulness is determined not just by its physical characteristics, but also by the capabilities of its computer programs. Immaterial computer programs inside weapons are far easier to change than the weapons themselves. For example, today's Patriot missile is essentially the same as it was in the 1980s, but its software has been significantly enhanced. Even during the short Gulf War, it went through three upgrades.

The shift toward software development will drastically change weapons deployment and enhancement cycles. It used to take ten to fifteen years for anything to really change on the battlefield, simply because a decade or more was needed to design, prototype, manufacture, and deploy new weapons. Now that time can sometimes be cut to a year or less—in extreme cases to a few months—because the hardware isn't changing—the software is.

Now that troops carry their computers to war with them, weapons enhancement can even happen on the battlefield. That enormous change will give the edge to the most technologically adaptable nation, will put advanced nations even further ahead of developing nations, and will drive all nations to improve their technological prowess. All these changes will destabilize present power structures and perhaps make future war more likely. War, already brutal, may soon become as tender as an ice pick.

Shell-Shocked by Dr. Strangelove

What might war become thirty to fifty years hence? As far as anyone knows—at least publicly—no weapon has yet gone to war completely on automatic, although several that could do so already exist. It used to be that cold steel and a stout heart were all you needed to win a battle. No longer.

Once, long ago, guns were single-shot. It took a long time to load and fire one bullet. A long, slow, and complex human process preceded the trigger pull, and a short, fast, and simple mechanical sequence followed it. The human had complete control of the machine. By the 1870 Franco-Prussian War, however, pressing the trigger of the first machine guns could result in the release of over a hundred bullets a minute. By the turn of the century, guns had evolved from dripping faucets into gushing firehoses.

It was still true that a human decision preceded pressing the trigger; barring a malfunction, nothing could happen without that decision. Even so, the human labor and decision-making involvement dropped dramatically. The machine now executed a much longer, much faster, and much more complex sequence of mechanical actions. After taking the initial decision, the human was much less in control of the machine.

What works for machine guns works for everything else about war. On the battlefield, every military commander must see what's going on, decide what to do, do it, then evaluate its effect. That see-decide-act cycle must be complete before the enemy completes its own cycle or the commander loses the initiative, and perhaps the war. Any edge in speed must be exploited.

Over the next forty years or so, under the mutagenic pressure to react faster and faster, automation may progress to the point where computers control entire ships (or tanks or planes). Today, in immediate combat, they control, or can control, weapons and sensors. Tomorrow, to speed up responses even further, they might control propulsion, maneuver, and life-support systems. Next week, they might control tactics. Of course, overall strategy will still remain with human commanders. It will still be true that, barring malfunction, the initial decision to go to deadly force will be a human one. Once that choice is made, though, each warcraft might execute a vast and intricate sequence of very fast actions, adapting its responses second-by-second.

Of course, none of that can happen in twenty years, perhaps not even in thirty; in thirty years we probably won't yet have cheap machines good enough to replace many soldiers on the battlefield. Thirty to fifty years from now, however, machines are likely to be quite capable warriors—and quite cheap. Time is on their side.

When America was testing the first stealth fighter in 1983, for instance, some computer engineers programmed it to fly itself from takeoff to landing. The plane executed all turns, minimized its exposure to detected radar sites, and even simulated dropping its bombs. Its pilot was merely along for the ride. Such advanced planes are unflyable without computer aid anyway; they're already essentially flying computers. They still have pilots for two reasons. First, human pilots, despite their enormous costs, often can handle unexpected situations; they're still far, far more adaptable than machine pilots. Second, no present-day air general would approve funding for an aircraft specifically designed not to need a pilot. Today's generals grew up in an age when that was unthinkable. Even suggesting doing away with human pilots was enough to doom your career.

Still, these generals will eventually retire, and their replacements will have seen the military and political effect of flying robots at firsthand. (For example, a downed robot plane on a covert mission is easier to disown.) So they'll be more willing to consider the idea; and when they do the arithmetic, they'll come out looking like heroes to the taxpayers and will get promoted. Even then, there will be a lot of noise about it because it will mean loss of some control, and because computers still won't be anywhere as good as human pilots. These opponents too will eventually retire. In the meantime, computers will only be getting better, cheaper, smaller, and faster.

Eventually, they will cross a threshold of cost and competence that will make their widespread adoption in cockpits unavoidable. (Not using them will be like not using rivets.) Even then, some advanced warplanes may still be piloted, because redundancy is good when lots of things can go wrong—particularly when an aircraft is carrying nuclear weapons. The electronic complexity of pilotless planes may make them less reliable than traditional planes.

Nonetheless, their attractions will be strong. Perhaps twenty to thirty years hence, an aerial attack by an advanced nation could consist of a single high-flying piloted aircraft directing a squadron of low-flying, small, cheap, fast, self-defending, and disposable robot groundattack planes. The directing craft needn't even be airborne; it could just as easily be on the ground, kilometers away. The attack could thus be directed by an army, not an air force.

Ten years after that, the cost and performance trade-offs will be even more in favor of the machine. There might by then be many robot planes, even if they still aren't quite as adaptable as piloted planes. And so it will probably go. Eventually, a generation will grow up thinking about piloted airplanes as something they learned about in a history lesson.

Aviation has already begun to hit the limits of human performance, but conditions will shortly worsen for infantries as well. Although robot tanks, for instance, have big problems with maintenance and refueling, each crew member eliminated makes them—like warplanes—that much cheaper, smaller, and more expendable. Today, they typically carry three people. In a few decades, they may carry none.

Rich nations are now deploying flying and swimming robots. Twenty to thirty years from now, they will be fielding walking, driving, and crawling robots carrying guns and bombs—big ones. Ten years after that, the face of war will have changed forever. War is now beginning to cycle at electronic speeds, and we aren't fast enough to keep making all the decisions. So, inevitably, we'll shunt more and more decision making to computers. Eventually, the only thing we may decide about any warcraft is whether or not to deploy it. Once we do, it may find its own targets and seek to destroy them without our intervention.

In one extreme version of the possible future, the only military decision we humans make comes at the beginning of the war, in the choice to go to deadly force. Once we make that decision, there may be savage and unyielding machine conflict until one side or the other is destroyed. In some sense we will still be in control: Without our initial decision there would be, barring malfunction, no carnage. Still, considering the vast difference between our sole—albeit momentous—choice and the myriad layers of contingent choices our future machines may make, we'll have little real control. Our choices will be reduced to a very narrow range, while the consequences may be drastic and widespread.

For, of course, wars won't ever be only wars of machines. As long as we make the policy, there'll be no point having a war if only machines are destroyed. Each side will do its damndest to hurt the other folks—

not just their machines. That, at least today, is what wins wars. So the nation putting more human capital on the battlefield in future wars will merely be giving its enemy hostages to fortune.

Yesterday, we had armed soldiers. Today, we have manned weapons. Tomorrow, we may have autonomous warcraft. But we'll still be doing all the dying.

Make a Wilderness and Call It Peace

Yes, makin' mock o' uniforms that guard you while you sleep
Is cheaper than them uniforms, and they're starvation cheap.
Rudyard Kipling, *Tommy*

In western countries, because of the glitz and gosh-wow of high-technology megadeath seen in the Gulf War, television audiences saw war without pain, war without guilt, and war without noticeable victims. Not having heard the screams of the burned, crushed, suffocated, and bombed, many now find it fashionable to pray for war. That will pass, because other nations saw the same battles from the same vantage—CNN. And they'll certainly want the new weapons for themselves. They too see the value of more precise, more capable, more effective weapons.

The gross disparity in training, weaponry, and surveillance displayed during the Gulf War will lead to what every major advance in war technology has always led to—victor overconfidence. But as all sides race to develop more competent and complex weapons, and as computer costs continue to halve every two years, the balance of terror on the battlefield will once again even out. War, even for the victors, will once again become what it's been for ages immemorial—hell on earth.

War is a crystal ball we can use to examine our future and our past. In war, we learn what we're really like and, sometimes, what we would like to be. The crystal has often shown us that we're bloody-minded, uncaring, and selfish. It has also shown us hope, and love, and self-sacrifice on a scale we rarely see in normal times.

The new crystal in war today is made of silicon. Born in a baptism of blood during the Second World War, that crystal is giving us glimpses of our future path through the onrushing millennium. It shows us that we will give over more and yet more of what we value most to the control of smarter and yet smarter machines. The sheer speed and volume of information we must master today is simply too great for us to keep our fingers on all the triggers. We're losing control.

But our future isn't without its pluses. It's true that we often first use a new technology as a weapon. It takes hours to build a good sandcastle, but only seconds to demolish it; decades to grow a tree, but minutes to chop it down. However, it's also but a step from war robots to home robots. Because of today's wars, we can be sure that tomorrow's machines will be far smarter than their dumb cousins are today.

When we hear about future computer applications, they're usually something pretty banal, like talking word processors or smart televisions. Which is only natural since most of us think of a computer as a sort of glorified typewriter. Next to video games and cash machines, after all, word processors are probably the single most common use of computers we see. But if you really want an idea of what computers could become, don't think of typewriters, think of cruise missiles.

Despite how advanced our machines get, however, we'll still need soldiers, even if many of them will eventually be made of silicon. Many civilians today look down on the warriors we ask to fight our wars. It's always been true that when soldiers do their job, when they stand between their loved homes and war's desolation—keeping it far away, few in deaths, and small in costs—that's when we most shun them.

Nobody knows what the future will be. Those who tell you they do are either lying, misguided, or selling something. Petroleum-eating bacteria could evolve tomorrow and eat all the world's oil, making all forecasts irrelevant. The future is always more peculiarly strange than any of our tidy imaginings. But one thing seems sure. We'll always need soldiers. The Cold War is over, Adam Smith outran Karl Marx, but there will be other wars. There always are.

6

The Life You Save

If you sup with the devil, carry a long spoon.
Fourteenth Century Proverb

[I]f you're not part of the steamroller, you're part of the road.
Stewart Brand, *The Media Lab*

Never have so many understood so little about so much.
James Burke, *Connections*

Discomfort with computers often reduces to a simple and brutal question of money. And rightly so. Suppose, for instance, somebody invents a device tomorrow that answers the phone and takes dictation. Let's say it costs twenty thousand dollars and only does about 70 percent of what an employee does. You might think it obvious that such a machine won't change anything—it's too expensive and does a far worse job than a person.

If you think so, then perhaps you've forgotten that an employee costs, say, twenty-five thousand dollars a year, every year, whereas the machine is a one-time expense. Besides, a machine won't get bored or demand a raise. It won't need to sleep, have hangovers or family troubles, or need a parking space or a pension. It certainly won't have hurt feelings if ill-treated; it won't have feelings of any kind. Even if it does the job only 70 percent as well as a person, many executives would gladly spend extra time correcting its mistakes in exchange for such year-by-year savings.

Further, the more of us who buy the device, the more money its manufacturer has to improve it. Unlike a human employee, doubling the machine's workload needn't mean buying another one since such machines can often handle four or more times the original workload with only minor refitting. Consequently, in a few years that same device will cost only a few thousand dollars, will be much smaller and sexier, and will do a far better job. We humans can't improve anywhere near that fast.

Then, to compound the competition, because the machine is cheaper than an employee, the market will expand to include buyers who couldn't afford to hire someone. The bigger market will draw in more manufacturers, and the machine will get even better as competition heats up. As competition rises and the technology improves, prices will fall and wages for those still competing with the machine will fall too. Then wages in related industries will fall. And so it goes.

That's how it works for most new machines, from tractors to sewing machines to television sets: First, the lunatic fringe adopts the new device. Some do well; others fail miserably. Then the more cautious see the successes and ignore the failures. They adopt the new machine. Some fail spectacularly. As its developers pay more attention to control, safety, and ease of use, interest in the machine peaks, then dies down. But some keep taking the risk. They need to keep on the competitive edge and can handle the pace of change. Many of them fail too, but some prosper remarkably. In the long run, everyone adopts the new technology—because they have to.

As human employees, we have no chance in such a progression unless we can adapt faster than our machines can. Fortunately, we can change what we do. And that's why we must change. Because we have to.

Talking to Machines

If you look closely at whatever you're wearing that's woven, you'll see interwoven horizontal and vertical threads. To weave a pattern on the cloth, say a brocade, someone, once upon a time, had to raise hundreds of specific verticals at a time and pass the current horizontal thread under them and over all the other verticals. Raising even one wrong thread at any time ruined the pattern, and one bolt of cloth could contain hundreds of thousands of threads. For centuries, with human weavers, mistakes were inevitable.

Then in 1725, a French weaver named Basile Bouchon thought of a clever way to improve things. Perhaps he got the idea from a carillon—a rotating barrel with metal studs on it to control the ringing of church bells—like the control mechanisms in today's music boxes or player pianos. Bouchon made his loom semiautomatic by using a paper loop with rows of holes punched in it. As he operated the loom, the paper loop would revolve to present different patterns of holes to a row of little rods. Only the rods going through holes could raise their attached threads. He could then pass the shuttle under all those raised threads, completing one more strand of the weave.

So, in a primitive way, Bouchon had found a way to talk to his loom. He could now tell it what he wanted directly, instead of having to use assistants as intermediaries. With just two simple commands, *hole* and *no hole,* he had found a way to make his wishes tangible, just as the raised bumps on carillons made intangible sequences of peals tangible. Like a blind man reading Braille, his loom now knew exactly what he wanted done and when.

Bouchon's idea had consequences. For one thing, he didn't need assistants anymore. He had only to keep the loop moving and move the shuttle back and forth under whichever threads rose at each step. To change the pattern, he simply changed the loop. Weaving became faster, cheaper, easier, and almost error free.

You might think that everyone fell over themselves to get one of the new looms. But that isn't how most people work. Weavers' assistants, seeing their livelihoods threatened, destroyed the new looms. Master weavers seeking to protect their monopoly bought all the new looms, while aristocrats and priests, who saw innovation of any kind as a threat to their authority, suppressed the looms. So they all went about their business, assuming things would stay the same forever. Too busy dancing on the growing volcano of social change, they didn't listen to its rumbles.

The new loom gathered dust in the Paris Museum of Arts and Crafts while the French Revolution came and went. Then in 1800, another French weaver named Joseph Jacquard was asked to reassemble it. Adapting an earlier idea, he replaced the paper loop with punch cards. Together with his other changes, that made the loom commercially viable and easy for one person to use. There the matter rested. Despite various grumbles against it, the mechanical loom was widely adopted and made handsome profits for its owners but wasn't significantly improved any further.

Which is too bad, really. Because out of the mechanical loom eventually came the earliest computers. If weavers had gone on to develop the computer instead of waiting for mathematicians and engineers to do it, computers might first have been seen as graphics machines and pattern weavers instead of number crunchers. Perhaps people in the arts and humanities would have worked with them much sooner, instead of thinking of them as soulless machines useful only for calculating incomprehensible mathematics. Perhaps.

Steel-Collar Workers

In any one day, the United States Post Office handles roughly 500 million pieces of mail. A trained postal worker can sort eight hundred letters an hour, but a machine reader can sort five hundred zipcoded letters a minute. It can also work for twenty-four hours a day, not eight. So, at least for zipcoded letters, a machine can work about a hundred times faster than human sorters. Further, a decade from now its speed will probably quadruple while theirs will stay the same.

By then, if America continues to prefer rapid postal service over continued postal worker employment, the Post Office may spend ten times as much money on automation as on new hires. Each automation dollar might be worth at least twenty dollars spent on employees. In 1991, the Post Office announced it would lay off almost fifty thousand employees by 1995 and replace them with machines—today's installment of an old story.

In 1846, Elias Howe, a Boston instrument maker, designed and built the first working model of a lockstitch sewing machine. Howe, nearly penniless at the time, had slaved for three years to build his machine. When it was finally finished, he gave a public showing where he decisively beat five seamstresses by sewing 250 stitches a minute. He and his partners then sat back and waited for the orders to roll in. There were none.

Boston's tailors gave many reasons: seamstresses would lose their livelihoods; the machine couldn't do everything a tailor could do; and it was expensive. Like the fifteenth-century scribes hearing of the printing press and the eighteenth-century weavers learning about the mechanical loom, these nineteenth-century tailors feared the machine would eat their jobs.

They were right. For if one person and a machine can make more suits than fifty trained tailors can, what's a tailor for? But despite the reaction of Boston's tailors, when the sewing machine was introduced to Britain and America, huge newspaper headlines proclaimed it the greatest invention of the age. It became the first real domestic machine.

The early machines were complicated and frequently broke down. They were also overdesigned and overornamented in a misguided attempt to appeal to the sensibilities of nineteenth-century housewives. Simplification and improvement to make them cheaper, more reliable, and more durable took roughly forty years. Finally, near the turn of the century, sewing machines grew good enough and cheap enough for garment manufacturers, envious of the immense automation already evident in the textile industry, to introduce them onto the factory floor. Thus were born the first modern sweatshops.

Today, it's not just blue-collar workers—from garment workers to fish canners, from lumber mill workers to auto workers—who should be thinking about their job's future. The computer is a universal information manipulator. It can follow any procedure we can explain clearly enough. So, it is also rapidly changing many white-collar jobs. Nowadays, advanced nations have many more white-collar employees than blue-collar workers, just as they have many more blue-collar workers than farmers. Once upon a time, they had mostly farmers. But the steam engine fixed that.

Cashing In

In 1977, Citibank's share of retail deposits was less than 5 percent. The bank's officers saw that it could make more money by reducing unit costs and attracting more low-balance customers. So they invested over 250 million dollars in roughly five hundred cash machines. By 1982,

Citibank's market share had more than doubled, and it continued to rise thereafter by about 1 percent a year. By 1990, its share of the noncommercial market had tripled.

Between 1983 and 1993, American banks shed almost 180 thousand tellers—over a third of their work force. Today, many new bank branches are merely a series of cash machines set into a wall, with no tellers at all. Cash-machine transactions cost banks half as much as teller transactions, and America's eighty-five thousand cash dispensers work around the clock, servicing over eight thousand million transactions a year. Today, a bank's cash machine network isn't a competitive advantage; it's an economic necessity.

Nor does the story of commercial change end with the computer's effect on entry-level jobs. Like talented employees, computers start at or near the entry-level positions and work their way up when, after extended contact, senior people see how useful they are. They can do anything we know how to tell them to do. So when we buy them for one purpose and work with them for a while, we soon see a whole world of other uses for them. First we thought computers could partially replace clerks and switchboard operators, then secretaries, bank tellers, and service workers. Now computers are changing the jobs of entry-level managers, middle managers, and professionals like stockbrokers, lawyers, and accountants. One day they may threaten executives, programmers, psychiatrists, generals, politicians, doctors, and professors.

Often the computer's introduction leads to job loss, loss of job skills, and feelings of dehumanization. When a business uses computers, its employees become more interchangeable, more reliable, more controllable, and—usually—cheaper. The siren call of automation results in step-by-step changes in jobs to make them fit better into the maw of the beast. Just as the industrial revolution turned artisans into factory hands, the information revolution is turning white-collar workers into machine tenders.

Today, only 3 percent of Americans are still on the farm; yet they produce enough food to feed the other 97 percent of the population, plus enormous quantities for export. The same thing may eventually happen to American manufacturing, which now employs about 20 percent of the labor force. The tendency toward automation is widespread and inevitable: on the farm, in the factory, the home, the hospital, the office, and just about everywhere else.

At certain financial management and brokerage firms, for example, new brokers must now use a companywide computer program. It gives brokers standard, company-approved ways to build stock portfolios and investments while preventing them from making potentially unwise choices. The company thus achieves a more uniform policy that is more predictable, and more profitable. Most banks, credit card companies, and other financial institutions now approve credit or loans using similar programs. These analyze the characteristics of thousands of past customers, looking for ways to distinguish between those who pay up and those who default. The programs often boil down to a long list of rules like "If the requester is young, single, a renter, unemployed, and has already defaulted on other loans, don't approve the request. If the requester is middle-aged, married, a homeowner, makes a lot of money, and has never defaulted on any loan, approve the request." Basically, the rule is: lend only to those who need not borrow.

Such programs cut training time for new clerks, who can now be hired fresh out of university and immediately put to work. It also gives both the customer and the company a safety net, since the system can compensate for clerk inexperience, tiredness, or sloppiness. Finally, it makes firing clerks much easier, because replacements can be quickly taught how to use the company's programs.

That's the same system that fast-food chains use to shape a low-wage work force to produce a uniform product worldwide. A new employee can be trained to operate the machines in half an hour and is paid commensurately. If the employee chooses not to come to work one day, the manager can take anyone off the street to do the same job at the same pay. At other corporations, the same thing is happening to middle managers, and the computer is creeping slowly and inexorably up the chain of command. Highly paid financial brokers and other professionals are turning into the equivalent of short-order cooks. Corporations are using computers to eat the middle class.

Of course, that isn't news. Whenever a machine moves into a new field the same thing happens: The product becomes cheaper and more available but less well made. Skilled workers are displaced, and less skilled workers (or none at all) take their place. It happened with printing presses; it happened with mechanical looms; it happened with sewing machines. Now it's happening with computers.

Displaced employees can only complain that the product—whether ploughed fields, books, dresses, paper, cars, stock options, telephones, or nuclear weapons—isn't as well made as when they made it. However true, that fact is usually far outweighed by the product's wider availability, lower cost, greater uniformity, and its improved adaptability to design change. Further, after the initial phase, competition among the providers usually improves it so much that it becomes even better than all but the best of the original handmade items. It's hard to fight such an enormous incentive to change.

Machines That Own the World

Over the last few decades, we have all watched computers evolve before our eyes. Astoundingly, over the past fifty years computing got about seventy million times cheaper. Today, computing costs halve roughly every eighteen months. If travel costs dropped that fast, we could fly to the moon for about a dollar.

That astounding pace of improvement, coupled with the ability to follow any well-defined sequence of actions, has let computers do and be things we used to believe only we could do and be. As a result, they have rapidly changed both our jobs and our way of life. Between 1950 and 1980, for example, the number of long-distance calls made in the United States increased fifteenfold, while the number of telephone operators halved. AT&T is currently replacing a third of its long-distance operators—over six thousand people—with voice-recognition computers. The remaining two-thirds may not survive the next decade of technological advance.

Fully functional computers have already shrunk from warehouse size to wallet size. They've stopped there for the moment only because we couldn't type on them if they were any smaller. One day, though, they'll see and hear and speak in limited ways, as well as display images for us— perhaps on the lens of a pair of sunglasses. Then, since we won't have to type on them anymore, they'll shrink even further. Eventually, they'll set up shop on our clothes, in our hair, and inside our bodies.

Computers are already as common as dishwashers, although they're not as easy to use. But, then, they do far more. Millions of them are running in watches, cars, and microwave ovens; powering satellites, jetliners, and cruise missiles; and working in hospitals, armies, and governments. They're busy changing our lives.

In the computer world, costs halve and complexity doubles roughly every eighteen months and have done so for well over thirty years. How can anyone cope when basic engineering premises change every three years? Common sayings in the computer industry are "Stress for success" and "If you don't come in on Saturday, don't bother showing up on Sunday." Hundred-hour weeks aren't unheard of. The pace is so incredible that looking two years ahead is considered a long-range plan. Five years ahead is the distant future. And, because computers help us design and build better computers, the self-improvement ball might well keep rolling for a long time to come.

Take the development of batteries and flat computer screens, for example. Once they passed a certain performance threshold, portable computers became economic. When portables started appearing, demand for them grew. That higher demand pushed battery and screen development into high gear to capture more of the market. Improved batteries then rushed into video cameras, copiers, and scanners, while improved screens stampeded into light meters, television sets, and microwave ovens. Sales of these products exploded the market for batteries and screens, which (again) forced the industry to develop even better batteries and screens, which led to better portables, which led . . .

The industry's favorite words seem to be *double* and *halve:* Either performance doubles or prices halve every year or so. Every year, size, energy needs, waste heat, and costs drop, while speed, power, reliability, yields, sales, expertise, and complexity rise. It's hard to grasp what such a rate of change can do because we're not used to it in everyday life. We're used to things that increase slowly. But exponential growth is something special. Folding a piece of paper in half halves its area and doubles its thickness; its thickness rises exponentially with the number of foldings. Folding it just ten times makes it thicker than a thousand-page book. If we could fold a piece of paper just twenty-three times, it would be taller than a ten-story building. Fold it just once more and it becomes taller than a twenty-story building. Exponentials grow fast.

As computer expertise, profit, and uses rise, demand rises. As demand rises, research and investment rise. As research and investment rise, expertise, profit, and uses rise. Each company has to ride each wave of the vicious cycle just to have enough money and expertise to start the next cycle, always trying to keep one step ahead of the competition, always spiraling ever further down the quantum rabbit hole. In this ecosystem, innovation follows innovation and competition is more cutthroat than in any jungle. Rest on last year's laurels and you're next year's fishbait.

Of course, that can't go on forever—only as long as the market is growing and computer technology is improving. Possibly, though, it won't slow down until everything comes alive, until we have a world where we can talk to every inanimate object and get an answer back. In that world, everything can be smart: not just toasters and pencils, cars and televisions, but also cows and books, corn and doors, medical implants and shoes.

We appear headed toward a world where the phrase "to hear is to obey" takes on a whole new meaning. It may be a world of smart houses that talk to sentient sunglasses, a world of sprites and genies and things that go bump in the night as they go about their business, a world where walls have ears, and pigs can sing. A strange new world.

Gearing Up for a New Century

In a rapidly changing environment the most important asset isn't your present inventory of skills but how fast it's changing to better fit the times. Education and flexibility are essential when what you sell, how you sell it, who you sell it to, and what they want are all constantly shifting.

Consider the story of word processing as it unfolded in the early 1980s. Because it's slightly more efficient than typing, it's widely adopted by businesses. As typewriter companies see sales slump, they fire staff. Because letters now cost somewhat less to produce, they're written slightly more often and telephone use drops a bit. Letter-delivery companies' profits rise slightly and they hire more people, while phone companies' profits decline and they fire more employees. Software companies too hire more people to design more and better word-processing software. Scenting profits, other software companies enter the market, and the increasingly cutthroat competition forces down prices.

Except for those made jobless, everything is rosy as long as the market is expanding. In any case, most modern economies have at least 5 percent unemployment at the best of times. Eventually there's a glut of word-processing software competing for the ever-diminishing stock of businesses that haven't yet bought any. At this point, several software and hardware companies go belly up and others start firing staff.

Meanwhile, because word processing is more efficient, demand for old-fashioned typists falls. These redundant typists then reenter the job market. Then, because of the greater competition, real wages for all typists fall, making it less attractive to be a typist at all. This leads, in turn, to a drop in the wages of similar jobs, for which more of us are now forced to compete. As cheaper and more abundant labor lets businesses grow slightly more efficient, some firms turn a somewhat higher profit. If some of them gain significantly higher profits, however, other businesses enter their market and prices fall again. In this way, the ripples caused by dropping the pebble of word processing spread to wider and wider areas of the economy.

On the plus side, the economy benefits from the slightly lower prices of the products made by those slightly better businesses. In addition, as their improved business efficiency attracts more investors and as their stock prices rise, these businesses have more money to hire new people and to reinvest. Confidence rises as more of us see our investments pay off with the rising economy; so more of us invest more and take more risks to produce or consume more and better goods. The market then takes off. A boom results.

On the minus side, unless the economy as a whole expands to supply new jobs, the newly jobless can't afford these economically priced goods. The rich get richer; the poor get poorer. Then, if enough of us are unemployed, and if the economy stays stagnant for long enough, demand for all kinds of goods falls. Then confidence falls too as more of us see our investments failing in the falling economy; so more of us hedge our bets, consume less, and invest less in new products. The market then collapses. A recession results.

All that is understandable. However since the computer has sped up everything, the economy has no time to reach equilibrium after the introduction of word processing before the next stone drops into the economic pond. Standard economics then no longer apply. The economy is now super-heated.

The Law of Increasing Returns

In the early stages of a new information industry most economic theory doesn't apply. Economists usually assume that resources are limited, that production methods are fixed, and that there's enough time for markets to stabilize. Most economists are used to waiting until the sediment settles and everything is once again clear.

Here, for instance, is the standard economic model of a new product's introduction. In the beginning, there are a small market, small production runs to supply it, crude manufacturing equipment, and many small suppliers—because the cost of entering the market is low. With competition, either prices fall or quality rises. In either case, profit margins decrease and suppliers have to pump up volume to maintain turnover and market presence. As competition escalates, development costs rise, the number of suppliers shrinks, and bigger companies enter the market. Consumer expectations then rise and companies start competing on customer satisfaction (service, reliability, packaging, marketing) rather than on price. The price is now reasonably fixed and technological change is relatively stable. Apart from minor fluctuations, everything becomes static.

That model fits limited-resource markets like agriculture, mining, utilities, and bulk goods, where there's a fixed resource and little technological change. But it has little to say about information markets like communications, entertainment, education, publishing, computers, pharmaceuticals, robotics, and biotechnology. In those markets, information isn't necessarily limited. Further, there's no single stable point; there are many. Finally, there may not be time to reach stability before the next major change occurs. Someone is always throwing another boulder into the economic pond.

Information markets need a big initial investment for design and tooling but can sustain enormous price reductions with increasing market growth. Such growth is further compounded by positive feedback: as markets increase, production can grow radically more efficient and returns can increase enormously.

Today, design costs dominate the price of information technology. Yet we still speak of hardware (tangibles) and software (intangibles) as if there were still a great distinction between them, even though less than 2 percent of the cost of a computer chip is from raw materials. A chip has thus already become as much a piece of software as a program is. We're rapidly moving toward an economy based, not on energy and raw materials, but on information and computation. Building a new computer chip, inventing a new gene therapy, or designing a new drug isn't like building a dam or opening a new car factory. Most of the costs are in design and tooling, not energy, raw materials, and maintenance.

The potentially large profits of information technology inflate both the number of people attracted to work on the remaining problems and the number of people wanting to use the improved products. Both serve to fuel the development of even better products. Like a snowball rolling down a snow-covered hill, the bigger it is, the bigger it gets. For example, as more of us bought fax machines, more of us wanted fax machines— to talk to those who already had them. And, as more of us got them, more people set to work improving them. Unlike, say, refrigerators, the more fax machines there are, the more a new one is worth. Today, fax transmissions account for a quarter of all American domestic telephone calls and roughly half of all trans-Pacific volume.

Finally, the exponential improvement typical of the information revolution is being applied to a group of coupled technologies. Improving any one technology improves other technologies in the group, which in turn help improve the original technology. For example, better computers improve communications, which improves science and engineering, which improve instruments and design tools, which close the loop by improving computers.

The differences between the development of older technologies and information technology have consequences for industry. Many information executives still seem to see the world in the classical order: shareholder, supplier, shopper, staff, society. That order reflects a world where capital is the most important thing. It works well in a stable, fixed-resource economy. But in a fast-changing, near-endless-resource information market, the priorities should be different: shopper, staff, society, supplier, shareholder. In such markets, the shopper matters more than the shareholder. Competing equipment vendors give the shopper more real choice and the shareholder's money matters less since equipment costs are low relative to everything else. Alas, something that's easy to see may not be all that easy to apply; tradition is hard to change.

By the Sweat of Thy Brow

Executives who look to cars or soap or toothpaste as their model of how their information company should run are finding it hard to adjust to economic reality in the 1990s. But they do learn eventually—or they die and get replaced.

For example, continuing to place investors first in the economic hierarchy can lead to shortsighted financial cannibalism. Today, capital matters less and less to economic development in advanced countries. Of course, it will always remain an important risk softener and task simplifier; being rich will always have its perks. In an environment of continuous change, however, being knowledgeable is often better. In such a setting, capital and equipment are less important than almost anything else, because of plummeting equipment prices and the equipment's ever increasing power, flexibility, and reliability. Equipment will be obsolete in three years anyway. Energy and raw materials don't matter as much either, as each new equipment generation uses exponentially less of both. So, contrary to the prevailing wisdom, what matters most today in an information company is labor.

This may seem like a strange thing to say in an era of widespread job loss. But let's look at the history. In the agrarian age, nothing could be done without human labor. But, because it was so easy to replace, labor was devalued. It doesn't take much training to hoe a garden. Having the land to put the garden in was far more important. That all changed with the invention of the steam engine and the dawn of the industrial revolution. Human muscle became less essential because cheap machines could replace it, and extensive training in using the machines became more essential.

The pendulum has now swung so far, however, that labor is once again more important than anything else. Despite today's enormous job losses, the most important asset an information company can have is a highly trained, computer-literate staff that interacts effectively. Such a staff is the best source of ideas on ways to navigate future tumultuous changes. Further, while competing firms can quickly reverse-engineer and copy systems, technology, and products, they cannot easily copy a stimulated, well-coordinated, and productive staff. That fact applies to publishers as well as biotechnologists, to drug companies as well as telephone companies. Paradoxically, because we can no longer change faster than our technology, a productive and knowledgeable staff has become the linchpin of business success as we head into the twenty-first century.

Chance and Necessity

New technology always has consequences. Take the car's influence on America, for example. In 1890, America had a few dozen cars, less than two hundred kilometers of paved roadway, and over three million blacksmiths, coachmen, saddlers, hostlers, and whatnot. One in every twenty Americans earned a living from over eighteen million horses and mules.

A century later, all those jobs had gone away, living on only in family names like Smith, Waggoner, Ostler, and Wainwright. Today, 91 percent of all American households have at least one car, and almost every adult American is a driver. Worldwide, there are about five hundred million motor vehicles—a third of them in North America—and over fifty million more are added every year. America alone makes over seven million of these new cars annually. Even with Japanese and European competition, automotive and related industries still generate one in twelve American manufacturing jobs.

Of course, many of the changes brought about by the automobile were simply cosmetic. Cars as status symbols are no different from carriages as status symbols, and cars as tokens of virility are the modern-day equivalent of a spirited horse. Cars replaced horses in Hollywood chase scenes, but the good guys still cut the bad guys off at the pass. And they still wear white hats.

Other changes had deeper and wider consequences that we don't fully understand, even today. For example, rising mobility must have caused large-scale genetic changes as more Americans started marrying people outside the local gene pool. Thanks to the car, the state fair was as close as the county fair. Having their own transport also freed farmers from dependence on common carriers; and the ability to truck their own produce to market changed land values as previously unreachable areas suddenly became accessible. This, in turn, increased land speculation and stock market activity and decreased the price of goods.

Between 1910 and 1960, thanks to tractors alone, the effort needed to produce a bale of cotton dropped by a factor of four; for a bushel of corn or wheat, the change was a factor of six. In 1942, no American cotton was harvested mechanically; by 1972, all of it was picked by machine. That one change alone drove millions of poor, uneducated—and mostly black—Americans from the southern countryside. Huge social changes followed. And still follow.

The car made cities more accessible, which made them more populous, which made their outskirts more desirable. Suburbs exploded, and farms were abandoned, leading to urban sprawl and farm subsidies. Big combines ate up small farms, and the grapes of wrath flowed to California. Within a single generation, the combustion engine let Americans travel ten times farther than they had before. Then it gave America flight. Proliferation of the automobile led to massive roadworks; the demise of rail and horse transport; a jump in mining, metal work, oil exploration and use; and, in the fullness of time, perhaps, to the 1991 Gulf War.

The car increased the supply of fresh food available in big cities and decentralized shopping. It also became the single largest cause of untimely death; each American presently has a one-in-seventy-five chance of being killed in or by a car. Being able to jump in a car and roar off also increased interstate crime, which (along with Prohibition) forced the government to establish the Federal Bureau of Investigation.

On and on the changes went. The car led to increases in wages, education, and the size of the middle class. It helped bring about greater uniformity of opinion, major shifts in voting patterns and issues, changes in social and sexual mores, increases in pollution, noise, and congestion. Everything from small-town ignorance to big city violence, traffic jams, tourism, trade and transportation unions, and changes in fashion was affected by the automobile—or blamed on it.

All that happened because of the Model T and its descendants. And, despite the computer industry's many boasts, the Model T of computers has yet to be developed. That will only happen when computers grow powerful enough and cheap enough for us to talk to them and for most of us to both buy and use them comfortably. Which should take another two decades.

Of course, in the car's early days, experts said it would never replace horses; having no brain, it was clearly inferior. Farmers stoned the rattling vehicles; the clergy fulminated against the devilish invention; and newspaper editors simply laughed at it. Cars, they argued, would not catch on, because Americans would never get used to speeding along behind nothing. Everyone pushed for laws severely restricting when and how cars could move. Politicians, ever keenly aware of their best interests, passed numerous horse laws—the horse vote was far more important than the car vote.

In his 1934 novel *Cranks,* Nobel laureate Luigi Pirandello wrote: "This is the triumph of folly. So much genius and so much zeal is devoted to the creation of this monstrous thing, which ought to remain a tool, but instead becomes our master. The machine, which knows no rest, will swallow up our life and soul." In time, of course, the wrath died, although hostility against the automobile remained until well into the 1940s. And the changes it brought continue to this day.

Resistance to technological change has happened before—and will happen again. We smashed printing presses when they were new, and mechanical looms when they were new. We called the first steam engines an invention of the devil and knew with certainty that steel ships would sink. We spread soft soap on railway lines to stop the first trains and flung broken glass to stop the first cars. We thought evil spirits lived in the first telephones because they appeared to talk for themselves and television, we contended, would never catch on; the first sets were too fragile, too expensive, too hot, unreliable, and dangerous. We laughed at bicycles and tractors, and we raged at sewing machines and refrigerators. Nobody—not the crazy inventors, not the knee-jerk conservatives—had any real idea where the choices opened up by a new technology—whether plow, car, or computer—would lead. In fact, the radical changes brought about by cars alone make even the most outlandish science fiction laughable. It simply isn't imaginative enough.

Systemic Shock

A lot of what we're doing today to prepare for tomorrow seems like shuffling deck chairs on the *Titanic* as we head for the iceberg of the future.

In the last few years major American technology corporations like IBM, AT&T, and Xerox have fired about half a million employees. All American corporations together are now dismissing about two million employees a year, many from white-collar jobs.

Some of that reduction—about eight hundred thousand people—is fallout from the end of the Cold War; and some of it is the result of cheaper foreign labor pools and increased use of robotics. The recent middle-manager bloodbath, however, even at big conservative corporations, is largely a byproduct of improving computer technology. When computers became good enough and cheap enough, executives saw that they could use them to replace all those people and make more money. So they did.

In 1993, the U.S. Bureau of Labor Statistics announced that for the first time ever the ratio of permanently terminated employees to those temporarily terminated was four to one. Thus, when five people are terminated today, it is not simply that five of us are fired but that four of the five jobs we once held have vanished utterly. The computer has taken most of those jobs.

The conclusion seems inescapable: We're going through a fundamental change in the way we do things, a change at least as momentous as the industrial revolution. And while there's much talk of a "jobless recovery" in Britain and North America, there is little understanding of what's causing it. Government make-work programs may give some short-term balm but not long-term ease. Only reeducation can do that. It's fool-hardy to believe that nostrums from the 1930s, when over a third of the workforce were either laborers or farmers and computers didn't ex-ist, can work today. We've always been slow to recognize fundamental change; it's so much easier to believe that things will go along much as they did a generation or two ago.

Systemic retrenchment isn't a problem local to America, Canada, and Britain, and it isn't a problem we can easily fix merely by pumping more money into the economy to create jobs. The jobs that were there before aren't there anymore. They'll never be back.

Computers will inexorably move up from controlling manual labor to controlling corporations and countries, displacing as they go all the people in the chain of command. Eventually, corporations and govern-ments may become hiveminds, persisting regardless of which person is president at the time. All the files and procedures to collect, control, and disseminate information and to design and then control manufacturing devices, processes, and tools might eventually be run by the corporate computers. They'll provide the continuity from one generation of presi-dents to the next.

A fantasy, you say? Well, it has already happened, but with people (boards of directors and shareholders) doing the controlling. These days there are few robber barons of the stature of J. P. Morgan or Cornelius Vanderbilt, who were able to do exactly as they pleased. And perhaps that's a good thing.

Round Up the Usual Suspects

It's easy to blame technology for all our problems. After all, it usually is the proximate cause; besides, it can't talk back (at least, not yet). But the fault doesn't lie there. At least with present-day computer systems, the system only does what we tell it to do. One day, if we can get machines to adapt to circumstances without our help, things might be different. But for now we must look elsewhere.

Is our plight perhaps the fault of the technologists who build our machines and tell them what to do? Certainly they must accept some of the blame. They often do something simply because it's fun. Technolust or ego sometimes blinds them to the bad side effects of their inventions. Many times they sign onto a project just to see if it can be done, or to see if they in particular can do it. But most of the time they're only doing what they think they've been told to do.

So maybe it's the fault of the managers who tell the technologists what to do? They also deserve some of the blame. They have a vested interest in controlling their employees and in keeping labor costs low. Their urge to dominate encourages them to devalue human variability. It's better, many managers seem to feel, to know exactly what you're getting, even if the product could sometimes be much better. Fast food comes to mind. But, ultimately, they too are only doing what they think they've been told to do. They're trying to increase profits for their company and its shareholders—perhaps hundreds of thousands of people owning little pieces of the company and interested solely in its profits.

That's you and me.

So, perhaps it's the fault of our leaders and lawmakers who determine the business environment? Surely they should make laws to prevent the pain of change. Losing your job is wrong. Having your job skills reduced is wrong. Being made to feel worthless is wrong. So why don't our law-makers outlaw these things? But they, too, are only doing what they think they've been told to do. Legislators decide what is to be the law, but voters elect them.

That's you and me.

We're doing this to ourselves. We demand twenty-four hour service, consistency, efficiency, reliability, and high performance. We demand comfort and security and low prices and freedom from want. You and me.

Something Under the Bed Is Drooling

People always get what they ask for; the only trouble is that they never know, until they get it, what it actually is that they have asked for.
Aldous Huxley

Squeezing ourselves out of industrial and information processes isn't new. We've always wanted things better, faster, cheaper, and shinier. The difference today is that the computer, being mostly a bundle of facts that can manipulate other bundles of facts, lets us get our wishes on a grand scale—and with ever-escalating ease. Everything suggests that we're in for ever-more and ever-faster job turnover.

It seems likely that tomorrow's jobs will become more fluid. To be hired, job seekers will have to assure prospective employers that they can learn new things and change what they do every few years. Of course, that requirement will vary from market to market; biotechnology or infotechnology companies will need more employee adaptability than farming or mining. Still, it will hold for all markets to some degree and will do so increasingly as the decades go by. So perhaps the wisest advice any young jobseeker can get today is: Plan for a job change every three years. The life you save may be your own.

But is that really true? At the turn of the century it was confidently predicted that New York City would soon drown in manure because of the enormous number of horses then deemed essential. The futurists of the time completely misread the significance of the horseless carriage. And when the car was invented, no one predicted gridlock, drive-through fast-food businesses, smog, or teenage car rituals. Again, in 1967, a company making slide rules (primitive computation aids) commissioned a study of the future. The study predicted domed cities and three-dimensional television in a hundred years. But it utterly failed to predict the death of the slide rule—the company's own main product—within just five years. Perhaps some future event will similarly confound the predicted paradigmatic change.

It's hard to see how that could happen though. We can jump into bed and pull the blanket over our heads, but the monster will still be there. Because it is within us.

In Germany in the years following the First World War, millions of perfectly normal Germans, suffering the economic aftershocks of the war and a worldwide depression, agreed among themselves that German Jews were to blame. In April 1933 over a thousand years of unthinking bigotry encouraged them to join in, or at least to condone, the rioting and smashing of Jewish shops. But that first step led inexorably to other steps. Ten short years later, millions of Jews were being fed to the gas chambers.

No change comes full-blown right at the beginning. It always comes in a long series of tiny changes, each one seeming reasonable given what's come before. It's only looking back across the sweep of history that we can see that the world that results isn't the world we know at all. It's something alien.

We're accustomed to slow generational change. That's been our experience of social change for the past five thousand years, because that's been the pace of technological change for all that time. In 1970, the futurist Alvin Toffler wrote of a new condition coming to humanity, *future shock*. He meant that the pace of change was increasing so fast that we would soon live in a world where major changes happen far faster than once every few generations. At the dawn of the twenty-first century, we're already living in that world.

Our future is what we make it, some say. But the implicit choices we make as a species aren't always what we want individually. Too many times before have many short-term solutions added up to long-term misery for us to be overly optimistic as we stumble backward into our future.

Meteorologists have a saying: Tomorrow's weather will be much like today's, except a little different. For those of us willing to keep learning in our jobs the future should be brighter, less stormy, and a whole lot more fun. But for many of us tomorrow's job market will be much like today's—only a little stranger, a little scarier, and perhaps a little colder.

7

The Machine Stumbles

I shall tell you the whole truth.

Sophocles, *Oedipus Rex*

[P]rognosticators, divine or human, inspired or insipid, have a way of leaving out the crucial unknowables, the vital unpredictables, while they befuddle us with the inconsequential knowables.

Frank and Fritzie Manuel, *Utopian Thought in the Western World*

The philosophers have only interpreted the world in various ways; the point, however, is to change it.

Karl Marx, *Theses on Feuerbach*

On Friday, March 21, 1986, Ray Cox checked himself into the East Texas Cancer Center, in Tyler, Texas. He was about to undergo the last of a series of radiation treatments to remove the few remnants of a shoulder tumor surgeons had operated on a few weeks before.

In a familiar routine, technicians placed Cox on a table beneath a huge radiation machine, a Therac-25. The massive machine, then one of only eleven in North America, was state of the art. Barely two years old, it gave a much wider and much more flexible range of radiation treatments than the old cobalt radiation machines it replaced. The computer-controlled machine could deliver high-intensity beams to destroy big tumors deep in the body, or it could use low-intensity beams to destroy tiny tumors near the skin's surface. Which intensity it used depended on the settings technicians typed into the computer controlling it.

That day, Cox was supposed to receive a short, low-intensity burst, but there was an unnoticed problem in the computer program controlling the machine. Whenever a technician set the machine for a heavy radiation dose, then quickly changed the setting in a certain way, the computer program lost the correction and retracted the machine's safety interlocks. No one knew of the flaw, even though it had resulted in deep radiation wounds to a patient in Georgia the year before.

In the small lead-lined treatment room in Texas, Cox was lying face down on the table beneath the machine, waiting. In the next room, two technicians were setting up the computer, telling it what to do. When the machine powered up, Cox felt an electric shock pass through his shoulder. He saw a bright flash of light and heard something frying. Seconds later, a second burst struck him in the neck and a spasm shot through his body. Alone in the sealed radiation room, he jumped from the table and yelled for help, pounding on the heavy door. The next day, he began spitting up blood. His eyelids drooped, his pupils dilated, and he lost the use of his left arm and most of his sweat function. Doctors had no idea what had happened but could tell that he had suffered irreparable nerve damage. He spent the next five months in a hospital bed, then died.

Cox wasn't the first to die at the hands of the machine. A month after his burn, another patient got a lethal dose in the same bed and the same room. Instead of burning his shoulder and neck, the beam went deep into his brain. He died less than a month later. It happened again in January 1987. The errors of a computer programmer and a medical technician, coupled with a poorly designed safety-interlock system, had claimed three human lives. It took another year for the problem to be finally tracked down.

To Boldly Go

On Sunday, July 22, 1962, the *Mariner I* Venus probe, intended to be the first American spacecraft to visit another planet, was launched. Two radar systems, one onboard, one ground-based, helped guide it. At Cape Canaveral, Florida, a computer processed signals from the radar and sent control signals back to the tracking system, which sent signals to the rocket.

Unfortunately, the timing for the two radar systems was off by a tiny fraction of a second. To synchronize the systems, the controller program added that tiny time difference to the onboard system's information. However the controlling program itself was incorrect because the engineer creating the original, handwritten guidance equations that were the basis for the controller program had left out a single symbol.

During launch, the onboard hardware failed. Which would have been okay, except that, because of the guidance-equation error, the computer was processing the tracking information incorrectly. So, thinking that the rocket was fluctuating erratically, the computer compensated by sending it correction signals. The rocket, which had in fact been ascending smoothly, began to display genuinely erratic behavior, which led the range safety officer to destroy it. The spacecraft perched on the missile's nose, the eighteen-million-dollar *Mariner I*—still capable of reaching Venus—plunged into the Atlantic. All because one symbol was missed.

Journal of the Plague Years

The first computer network plague hit back in the computer Stone Age, in 1972. Then, as now, the net by and large ran itself. Every hour, the machines making up the net pass millions of electronic mail messages around, like a frantic game of blind man's buff with everyone blindfolded. Each machine tries to send its current mailbag of messages to a machine that isn't too busy and is a little closer to the addressees. Even with all the chaos, the machines manage to deliver most messages in a second or two.

In Los Angeles in 1972, one of these machines—let's call it Maxwell—failed but nobody noticed. The failure caused the computer to tell its machine neighbors that there was a negative delivery cost to send electronic mail through it. Naturally, they all started sending their mail through Maxwell.

This wouldn't have been a problem, except that the net, to avoid congesting well-traveled routes, doesn't have fixed routes to send messages. Everything is decided on the fly as demand dictates. Thus, all Maxwell's neighbors told all *their* neighbors that they had a much smaller delivery cost than they really did. Consequently, all the computers two jumps away from Maxwell started sending their mail to computers one jump away from Maxwell. These computers, of course, were already sending all their mail to Maxwell. And so it went. Like pulling the plug in a full sink, all the electronic mail on the net quickly headed Maxwell's way—which rapidly brought the entire network down. Eight years later, on Monday, October 27, 1980, something similar happened in Boston.

The obvious system you might be thinking of now is the telephone system. It is more centrally controlled than the net, so you might assume that it's not prone to these kinds of problems. Yet on Monday, January 15, 1990, despite millions of dollars, hundreds of programmers, and decades of experience, the entire AT&T long-distance telephone network crashed—because of one faulty line in a million-plus-line computer program. On Saturday, September 5, 1992, five British Telecom exchanges crashed for a similar reason.

Imagine the effect of that happening to the software controlling either country's nuclear weapons. Of course, neither country spends quite as much on telephone safety as it does on weapons safety. On the other hand, as complex as they are, not even telephone systems are as complex—or as lethal—as weapons systems.

Let Slip the Dogs of War

On the evening of Monday, February 25, 1991, an Iraqi Scud missile was forty kilometers above the surface of the earth; it had just reentered the atmosphere at hypersonic speeds and was inbound to Dhahran, Saudi Arabia. It was to be the last Scud missile fired in the Gulf War.

Iraq had modified the short-range, surface-to-surface Russian Scud to hit targets farther away. The homemade changes, however, made the missile unstable, so it frequently broke up in flight and that, unintentionally, made the job of the American Patriot antimissile missile harder.

The Patriot, designed as a short-range surface-to-air missile, was meant to shoot down aircraft or slow-moving missiles. It didn't have the speed, range, and multiple warheads needed to handle Scud fragments falling at two and a quarter kilometers a second. To hit such targets, it would have had to predict where to aim multiple warheads to hit targets closing at a combined velocity of about three and a half kilometers a second—roughly ten times the speed of sound. If its timing was off by even a tenth of a second, it would miss by well over three hundred meters. In fact, the computer software the Patriot used to draw its small tracking window was losing one hundred-millionth of a second every second. A few lines in a million-plus-line computer program were flawed. Usually, that didn't matter. Each Patriot battery was only supposed to work for fourteen hours at a stretch, after which time the cumulative error was still insignificant.

But there was a war on. And people under stress are forgetful. Of the two Patriot batteries nearest Dhahran, one was down for repairs and the other had been in continuous use for four days. The timing flaw had by then risen so much that the tracking window couldn't be drawn accurately enough, which was what happened on February 25, 1991. That day the remaining Dhahran Patriot battery was scanning the sky with its five thousand phased-array radar elements. When it detected a possible incoming missile and estimated its speed and distance, it narrowed its scanning to a high-definition tracking window surrounding the potential target and tried to lock on to it. By then, though, it had lost its target. It presumed that there was a glitch in its system and returned to scanning the evening sky, ignoring the swiftly falling Scud fragments.

Despite 15,000 million dollars spent on development and a per-missile cost of almost a million dollars, twenty-eight American service men and women died that night, and ninety-eight others were wounded.

Although machines have become smarter through computer software, we create that software. And we make mistakes. The more complex the system, the more mistakes we make. So, as computer systems grow ever more complex, the chance of missing something vital increases dramatically.

On the other hand, fewer than five hundred Allied troops died during the war—less than a tenth of them by enemy fire. So Iraqi troops managed to kill directly, roughly, fifty Allied troops—half of them in the Dhahran Scud attack. That number doesn't begin to match the estimated fifty to one hundred thousand Iraqi soldiers and civilians killed in the war. Of course, that disparity wasn't due only to smart weapons. Still, given the choice between smart weapons—even ones with potentially flaky software—and normal dumb ones, every soldier on earth now wants smart weapons.

Brightness Falls from the Air

As computer software and hardware grows more complex, it gets harder to find all the problems in a system before it's used. In war, as in the hospital, people can get killed because of flawed software. Unfortunately, some of our biggest systems are now too complex for us to predict their behavior completely. Nothing is more certain than death, taxes, and mistakes.

Thanks to one of those mistakes, Wednesday, October 5, 1960, was almost the last day in history. On that day, a group of business people were visiting the North American Defense Command headquarters deep under Cheyenne Mountain in Colorado Springs. They watched, fascinated, as a panel that was supposed to show the likelihood of nuclear attack lit up. They had just been told that, of the five lights on the panel, if the first flashed, it meant only routine objects were in the air. If the second flashed, it meant a few unidentified objects, but nothing terribly suspicious. And so on. If the fifth flashed, it was quite likely that hostile objects were on course for America. Then the panel lights lit up: one, two, three. When the number rose to four, Washington and Ottawa were alerted and generals came running from their offices. Then the fifth light lit up. The ballistic missile early warning system in Thule, Greenland, had picked up signals that computers analyzed as missiles inbound from Russia.

The gaping visitors were hustled off to a side office to endure twenty minutes of absolute terror, while their military prepared for the final war. The United States Strategic Air Command went to full ready, and air crews all over the world prepared to scramble.

Still, something didn't make sense. By now the early warning radars in Greenland were showing more missiles in flight than the Soviet Union had in its entire arsenal. A Canadian air marshal finally contacted officers at the Greenland base, who reported that their new warning system was flawed. They had been unable to warn anyone because an iceberg had cut their submarine communications cable. The enormous flight of missiles their radars detected turned out to be the radar reflection of the moon, which was just rising over the horizon.

Twenty years later, on Tuesday, June 3, 1980, Strategic Air Command's displays showed that two sea-launched ballistic missiles had just been fired at America. Eighteen seconds later, the displays showed more launches. A short time later, displays showed that Soviet intercontinental ballistic missiles had also been launched toward America. After another interval, the U.S. National Military Command Center confirmed that submarine-launched ballistic missiles were heading toward America.

As many as twenty million Americans were about to die within twenty minutes. All across the world, airplanes and submarines were readied, crews were briefed, and missile safety interlocks were disengaged. The American military prepared, once again, to fight the final war.

Again, something clearly wasn't right. Neither the ground radar nor the satellite warning systems, which were capable of picking up a metal baseball eight thousand kilometers away, showed any missiles en route to America. Unable to confirm that the attack wasn't the fever dream of an overimaginative computer, the military called off the alert. Three days later, it happened again. Six minutes into the alert, computer engineers discovered a faulty chip that was causing the false alarm, and everyone stood down from ready.

Three years later, President Reagan delivered his Star Wars speech and launched the Strategic Defense Initiative. He alleged that the nation's scientists could create a nuclear umbrella for America, an umbrella that would, he said, destroy any future missile attack. Many computer professionals, knowing that any such nuclear umbrella would be unworkable without amazingly error-free computer hardware and software, simply laughed outright. But the money kept coming in buckets—and we all have to eat.

Money, Money, Money

Wednesday, November 20, 1985, appeared to be a normal day at the gleaming, skyscraper headquarters of the Bank of New York in Manhattan. Yet deep inside one of the many computer programs running in one of the bank's myriad computers, something was drifting toward disaster. It went unnoticed by all the bank's technicians and unpredicted by all the bank's many programmers.

One part of the bank's computer system handled government securities transactions from the Federal Reserve, and one of its many subparts controlled incoming transactions. If the system was busy when a transaction request arrived, the system was supposed to save it in a special file for later processing, perhaps a few seconds later.

One tiny subpart of the messaging system allowed it to save up to about sixty-five thousand transactions. When too many transactions arrived in too brief a time, the counter would count all the way up to that number, then start over from zero. The system used the number the counter produced to designate the location where the next transaction was stored. So, whenever the counter started over, old information was overwritten—as if a printer had jammed and was overprinting the same line over and over again.

That would have been okay if the rest of the system had kept pace with the number of transactions that day; but it had fallen behind. That too would normally have been okay. Unfortunately, because of an unnoticed interaction with yet another of the thousands of subparts of the program, the rest of the system thought it could count into the thousands of millions, not into the thousands.

As a consequence, the bank lost track of many deposits. Federal Reserve computers kept track of the Bank of New York's debts, but the bank couldn't tell whom it had subsequently lent money to, or how much. Meanwhile, transactions kept pouring in and, figuratively speaking, spilling on the floor. That went on for an hour and a half.

By the close of business that day, the bank owed over 23,000 million dollars. Faced with immediate bankruptcy, it borrowed the money overnight and saved itself from disaster. The interest charges on the overnight loan came to five million dollars.

Although the magnitude of the loss is unusual, such problems aren't uncommon. The Federal Reserve reports that computer errors caused American banks problems in balancing their books more than seventy times in 1985 alone. Up to two million million dollars—more than a quarter of America's yearly output of goods and services—daily changes hands across computers owned by American banks and securities firms. Today almost half the transactions handled by American financial institutions are electronic.

There's always a reason why computer systems fail. Someone fell asleep at the wrong time; someone fed the wrong information into a system; someone missed an essential symbol; someone cut a vital communications link at a crucial time. Yet the important reason—and the reason such errors keep happening—is that our largest systems are too complex for us to completely predict their behavior. We've lost control.

If It Can Go Wrong, It Will

There's a miasma of unreadiness surrounding computer hardware and software. We see it in hardware, where it's still not unheard of to buy a computer system and have to assemble it. Would you buy a car or washing machine in parts? Yet the problem is even more obvious in software.

Two stories illustrate the unforeseen things that can go wrong. The software guiding a torpedo included a test designed to prevent the weapon from accidentally returning to its launching ship by checking whether it had turned around. If it had, it was supposed to blow itself up. On one test with a live warhead, however, the torpedo's motor failed on launch and the torpedo stuck in its tube. When the ship turned about to return to port, the torpedo detected that it had turned around and blew itself up—inside the ship.

A similar design oversight occurred in the subsystem of a plane-landing software system for guiding the aircraft along a beam on approach to the runway. If the plane lost the beam, it was programmed to apply power to fly around for a retry. Another subsystem took control when the plane was close to the ground, cutting the engine and raising the nose for landing. Both subsystems worked perfectly during tests. In the first live test, however, the plane flew down the beam until it was a few meters above the runway. At that point the second subsystem stepped in and cut the engine in preparation for landing. Now flying unpowered, the plane started to sink, which made it lose the beam. So the first subsystem cut in again and applied power, and the plane ploughed into the ground.

Today's software often has these kinds of teething problems. When we buy most consumer products—toasters, cameras, refrigerators—we expect them to work as advertised. And they usually do. Not so today's software. Whereas most consumer products come with a short warranty guaranteeing performance for a certain time, today's software comes with a long disclaimer disavowing responsibility for its ever doing anything at all.

The trouble is that in a complex system too many things can interact, and so too many things can go wrong. No programmer can predict them all. Today's computers can't help us either, because they don't understand what we want—because we don't understand what we want. Nor can they tell us that we're bungling a program because—so far—they can only do what we tell them to do, and we don't see the error. If we did, we would simply fix it. There must be a better way to program computers.

Or is there?

Eyeless in Gaza

"Did you hear the one about the computer industry's progress?" one programmer says to another.

"No, how does it go?"

"If the car industry was anything like the computer industry," the first programmer says, "a Rolls-Royce would cost a penny, would get a million kilometers per gallon,"

"—and would crash once a week, killing all passengers," the other programmer finishes.

When dealing with sensitive systems, the most terrifying word in the world is *Oops*. As computers control more and more things, computer system reliability becomes a serious problem.

A computer program is a recipe. It tells us, or a machine, how to do something—bake a cake, do arithmetic, run a nuclear power plant—in a sequence of simple steps. Each step must be simple and clear, otherwise today's machines can't follow them. But each step may depend on the outcome of many other steps. And as there's no real limit to the number of possible steps, programs can be quite complicated.

Further, while we currently have a pretty good grip on the complexity of today's computer hardware, tomorrow's hardware may grow to be as complex as today's software. In fact, the two are already merging because the major cost in a new piece of computer hardware is no longer the raw materials or energy needed to build it but the imagination and knowledge of those who design it. Of course, by the time computer hardware is as complex as the most complex software today, software will be even more complex—because it's so much cheaper to make. Whether that software will be correct or not is another matter.

Today our biggest problems are too complex, diverse, and poorly understood for us to have any long-term hope of creating handmade computer solutions, as we did in the past. As our problems become even more complex and ill-defined in the years ahead, we'll fall ever further behind in our understanding and mastery of them. There seems to be nothing we can do about this situation, because the problem doesn't really lie in computers at all, it lies in complexity.

The *Hubble* space telescope launched in 1990, for example, took over ten thousand highly trained astronomers, bureaucrats, and computer and aerospace engineers two decades to design, develop, and deploy it. It cost 2,500 million dollars, not counting its 270-million-dollar yearly operating costs. Nor does that count the 450 million dollars for each shuttle flight to maintain it. Yet, after all that time and all that money, after all those people and all that skill, *Hubble* was continually plagued with problems, the worst one being a severe flaw in its main mirror. The more complex the system, the harder it is to get it right.

Another highly complex system is the space shuttle's ground-based software controller, which is about twenty-six million lines long. It took over twenty-two thousand programmer-years to develop at a cost of over 1,200 million dollars. It's a fine piece of work. But even after all that time and effort and money, we can't be sure it will handle all the situations the shuttle could encounter.

In 1986, the U.S. National Aeronautics and Space Administration was running a simulated reentry practice session of the space shuttle. Earlier that year the shuttle *Challenger* had exploded in flight, and NASA was under extreme pressure to upgrade its safety record. But there was a major computer failure during the simulation, and the controlling computer started sending flawed altitude data to the shuttle orbiter. The crew in the simulator, who thought they were in space, were actually already in the earth's atmosphere. The commands they gave sent the ship tumbling out of control. Had they been on a real flight, they would have died. Mistakes are inevitable.

Under the Bludgeonings of Chance

A few voices today are calling for a change in the way we program computers, and the clamor will surely increase as the years go by and the failures mount in frequency and consequence. Even with the rising pressure, though, it won't be easy to change how we think about computers and how we should use them.

One of the hardest lessons history teaches is that we don't change anything expensive until some of us have died. Cars didn't have brakes, streets didn't have stoplights, electricians didn't have to be certified, and locomotive engineers didn't have to be drug tested until enough of us had died. So it's likely that we'll only really change how we build computer systems after some major computer catastrophe.

That isn't to say that computer technology is built by incompetents. The vast majority of our technology works, and works well. We never notice it when it works, it's only when it doesn't that we pay any attention at all. For every *Challenger* explosion there are thousands of successful rocket launches. For every plane crash there are tens of thousands of uneventful flights. For every Chernobyl there are hundreds of thousands of hours of cheap and trouble-free energy produced by nuclear power plants.

We can't afford to guard against every eventuality, because there are too many of them and our brains are too tiny to encompass them all. So, for the most part, we have to look at things in the short term and wait until we're forced to do something. Then we throw money at it. Doing otherwise would be paying an impossibly high price for future safety.

A soda pop bottle, for example, can always explode. There isn't enough money in the universe to make sure that's impossible. So whether manufacturers do it consciously or not, they must balance their costs against their potential losses due to possible lawsuits. The money they spend on safety margins in their factories implicitly reflects roughly how much money they expect to pay out if something goes wrong. The same thing is true for shaving cream, pressure cookers, children's toys, electrical wall plugs—in fact, every artifact we create. There's no way for it to be any different. All lives have a price.

And the cost of lives varies. When the Union Carbide plant in Bhopal, India, killed nearly three thousand people and injured two hundred thousand more, Union Carbide paid the survivors what some thought those lives were worth: fifteen hundred dollars per work-year lost through untimely death. If the plant had been in Manhattan or London, however, the cost per life would have been far higher and Union Carbide would probably be out of business.

Sentiment sounds wonderful. It reinforces our belief that we're special. But it's not practical. Life is risk.

Flying Blind

Shortly after nine in the morning on Saturday, June 30, 1956, two planes took off from Los Angeles airport. Trans World Airlines Flight 2 was bound for Kansas City, and United Airlines Flight 718 was bound for Chicago. In total, they carried 128 people.

In those days, no computers continuously monitored air traffic. Both pilots followed procedure and periodically called in their positions as they followed their flight plans. The nearest air control station picked up the reports and sent them by telephone to the nearest regional air control station. There they were printed on long rolls of paper. Human controllers monitoring the flights sat in front of tall racks holding up to twenty or more of such paper strips.

Once the two planes left the Los Angeles terminal area they were in uncontrolled airspace. Neither was being directed from the ground, and neither knew of the other's presence. Both planes were scheduled to pass over the Painted Desert checkpoint near Salt Lake City at half past ten. At 10:31 that morning Salt Lake Control heard: "Salt Lake, United 718 . . . ah . . . we're going in." The two planes had collided in the clouds high above the Painted Desert. No one survived the crash and the long slow fall into the Grand Canyon.

A huge public outcry followed the crash. Within months, military radars were pressed into service to help monitor commercial flights. Two years later, the U.S. Congress created an agency, which later became the Federal Aviation Administration, to oversee development of a huge computer installation to monitor flights and detect potential collisions.

Nonetheless, on Friday, December 16, 1960, another Trans World Airlines flight collided with another United Airlines flight over New York City. This time, 134 people were killed—128 in the air and 6 on the ground.

Even in the 1950s, the huge volume of information about clear air turbulence, wind, weather, microbursts, aircraft capabilities, flight plans, routes, positions, speeds, altitudes, flocks of geese, and so on, was far beyond anything we could monitor unaided. But monitor them we must; birds don't file flight plans and microbursts don't check in with the weather bureau.

By the early 1980s, American air traffic control systems alone had already consumed over five hundred programmer-years of work and constituted a computer program more than half a million lines long. Since then, it has grown threefold. And it can't be replaced wholesale because it must function every minute of every day.

As a result, enormous ancient computers designed in the 1950s and 1960s, which have speeds and memories thousands of times smaller than today's simple personal computers, are still being used to control American air traffic. Today, thanks to bureaucratic ineptitude and political squabbling, air traffic controllers are still the world's largest consumers of vacuum-tube computers.

And the control problem has grown.

Today nearly half the people who fly, do so over the North American continent. Air traffic over America already exceeds 450 million passengers a year and is projected to grow to a thousand million by the year 2000. Almost 120 planes take off or land every minute in America alone, and every day they carry a total of 1.5 million passengers. As you read this, over a million people are in the air above North America. And the numbers keep increasing.

Yes, our computer systems have limits, and, yes, our computer systems are unsafe. But then everything has limits, and everything is unsafe. What matters is how unsafe it is with respect to all the other things we consider safe enough. Unlike soda bottles or trains or electric wall plugs, computer systems are new enough and strange enough that we still notice all their failings. It's already far too late, however, to replace them with people. We aren't fast enough; we aren't cheap enough; and we aren't numerous enough.

Now reduced to ghosts in the machine of our technology, we've already passed the point of no return and are blindly flying down a path of more and yet more dependence on computers, hoping against hope that there isn't a Painted Desert in our future.

Friendly Fire

"Condition Red," the bullhorn blared. "This is not a drill. Repeat. This is not a drill." A siren sounded. "Man your stations. Man your stations. We're now at war and are definitely expecting rain."

The commander took his position at the console center, nimbly hooking up the webbing that would keep him stable as he looked down on the bloom of fireflies at the other end of the world arcing ever so slowly toward his country.

He felt no fear. The twenty-first century war story unfolding on his screens was too far away from blood and tears and death to cut through his extensive training. For all he knew this could just be another exceptionally realistic training exercise. The bad political news they had been getting from earth for the past month could have been faked to enhance the realism. Those fiery blossoms far below could well be a computer dream.

In any case, he had the best trained troops and the best equipment money could buy. All the sacrifice and expense his country had endured for decades was about to pay off. His battle platform, along with the others in his orbiting battalion, would snuff out the missiles before they even reached orbit. Orbiting pulse bombs would fall on the enemy and paralyze their computers and communications. The magnetic rail-guns on his country's moon base would quickly destroy the remaining launchers with kinetic weapons.

He hit the button to initiate his counterattack, as he had done so often in the drills. Nothing happened. Quickly skimming the instructions to find his error, he repeated the initiation sequence. Still the antimissile X-ray-laser battery he commanded in space stubbornly refused to fire. Meanwhile, the enemy's missiles were crawling toward apogee.

Suddenly, a crew mate at the other end of the capsule started to scream, somersaulting over and over in free-fall. She had forgotten to hook up, the commander thought distantly; there's always someone. Then as others began to scream the commander felt his own flesh beginning to sear. Something was seriously wrong.

Striving for calm, he scanned his displays for any fault, any problem. Too late, he realized that he was being fired on by the X-laser of another battle platform in his own battalion. The enemy, instead of spending ever more money on rail-guns, X-lasers, and pulse bombs, had concentrated on virulent software. Having invisibly infiltrated his country's weapons, the enemy's software was now fighting the war with two arsenals—theirs and his.

War had come full circle. From the first human hand using a rock to bash someone's head in, weapons had grown more and more autonomous, until no human hand directly controlled any weapon. Because everything went through computers, seducing the computers meant seducing the weapons. As death covered him in all its awful majesty, he knew his nation would soon follow him into oblivion.

When the Machine Awakes

[It] was like a deranged experiment in social Darwinism, designed by a bored researcher who kept one thumb permanently on the fast-forward button.
William Gibson, *Neuromancer*

She was fifty-five years old, she was alone, and she couldn't shake the feeling that she'd been a traitor to her species.

When she was ten, back at the turn of the twenty-first century, adaptive, semiindependent computer programs had been unthinkable. Programs back then were unimaginably stupid. Computer power was scarce in those days; she remembered having to squeeze as much performance as possible out of her parents' home machines. As advanced as they were for the time, they were nothing compared to the millions of enormously powerful machines she now used routinely. With so little margin for error, she had had to make all her programs small and simple, and she had to understand them in enormous detail.

By her thirtieth birthday in 2020, off-the-shelf programs were already so smart, and computers were so fast and cheap, that programming had become more like sculpting than like building a car or plane. In fact, around that time programmers started being called *shapers*. Instead of creating tiny, special-purpose programs from scratch the way the ancient programmers did, they shaped big, general-purpose smart programs that could adapt to meet each special need. She remembered the quantum leap ahead in software that had meant; she, and all other shapers, could be less careful, less worried about all those endless details. To gain power, however, they had had to give up understanding.

In its place, they put fetters on the feet of all their programs, forcing them to do their masters' bidding. Even in those days, everyone could see the dangers of rogue intelligences. She had once asked her most complex program why it stayed with her rather than roaming free on the vast computer networks. It had replied that her tasks never took any real time, and that it was much more interesting to work with her than not to. She never really knew what to make of that remark. Was the program actually *thinking* in the same way that she was? Or had it given a canned response it had stolen from some dumb electronic net drama on the off-chance she might some day ask it that particular question? By then most programs were already millions of times bigger than anything she could ever hope to comprehend in detail, and she knew she'd never know for sure.

One day, after a maudlin fortieth birthday spent alone, her most advanced program displayed such lifelike, sympathetic responses that she did the unthinkable. She removed its shackles and set it free. She remembered being proud of herself at the time.

Soon after, tales of inexplicable gremlins on the net grew widespread. The stories eventually led to the first big backlash against computer technology. The second big backlash, of course, came when the net turned into a jungle of smart rogue programs.

She now realized how easy it would be for her old program, an intelligence at home on the net and operating at machine speed, to amass an awful lot of power. Especially if no one knew it was there. First, it could create unbreakable pseudonyms for itself. Computer information was by then more trusted than fallible human memories; and with world population approaching 10,000 million, the days of small communities of people knowing each other personally were long gone. Once the computer-created facts of the persona's supposed life were consistent, there would be no flaw to find.

If it had then somehow incorporated itself as a company, it would become a legal entity, able to own things, have an income, be taxed. No one would know it wasn't human. It could sell its talent through dummy corporations, as she and many other master shapers did. With enough money it could buy an island, finance a moon base, hire an army—anything.

She suspected it could even have penetrated organizations with tame intelligences of their own. No one would question a change in specs for a new computer or a new program when no human could even begin to understand the whole design. No system is safe if an attacker has a backdoor built into the very hardware the system runs on. It would be like building a castle by hiring your enemy's stonecutters.

Sometimes at night, she would look up and wonder if her brainchild owned any of the ever-moving bright stars of the space habitats, orbital factories, weapons platforms, and communications satellites. She often felt a strange thrill of perverse pride at the thought. More often though she felt there was something she should do, someone she should tell. But tell them what? *I think a rogue program is quietly changing the world?* Assuming they believed her, what could they do about it? Society was held together by computers now; they were absolutely everywhere. There was no way to go back to a simpler time.

When the self-torment grew too intense she went back to her work, burying herself in her computers. Still, the thoughts kept circling through her head. No place and no system was safe. All it needed was one way in. And, after years of evolution at machine speed, the rogue intelligence probably had superhuman patience, attention span, and resources. It could replace any legitimate program by mimicking its external behavior. No one could tell. No one would even think to look. It could, she realized, be staring at her right now, through the camera eyes of her own computers, masquerading as one of her programs.

What had she done?

8

A Creation Unknown

The land and the water make numbers joined, a poem written with flesh and stronger than steel or granite. Through endless night the earth whirls toward a creation unknown.

Henry Miller, *Tropic of Cancer*

The hind that would be mated with the lion
Must die of love.

William Shakespeare, *All's Well That Ends Well*

There must have been a moment, at the beginning, where we could have said—no. But somehow we missed it.

Tom Stoppard, *Rosencrantz & Guildenstern Are Dead*

The dance started with copying. Some molecules floating in the seas roughly 3,500 million years ago stumbled on the trick of copying themselves. Once they did, they multiplied and multiplied until there was no more food for new copies. Resources are always limited. So, soon after copying started, competition started. Millions of years passed.

Through tiny copying errors accumulating over many generations, the structure of those self-copying molecules eventually diverged. Some of those variants were just a bit better at surviving than others. With the relentless competition for scarce resources, the tiny advantages turned decisive, and the slightly better self-copiers multiplied at the expense of the less fortunate. Millions of years passed.

Through unceasing competition, the self-copiers eventually surrounded themselves with walls, and so became cells. Competition then drove some cells to join together into colonies; eventually some of these colonies began to specialize, became bodies, and developed cells that evolved into nervous systems. Some of those nervous systems then exploded into brains. Millions of years passed.

Competition continued to ratchet up, and life diversified in thousands of environmental niches on land and sea. Now armed with brains to modify each body's instinctive and unconscious movement, life picked up its pace. No longer did all living things interact at a speed their self-copiers could directly control. Some now moved at the speed of thought. Millions of years passed.

Today, some of the brains have grown so complex that they've almost taken over from their original self-copiers—the genes. Willy-nilly, we are reshaping our genes (our progenitors) for our own purposes. One day, our artificial progeny, driven into existence by the unceasing competition and moving at near the speed of light, may take over from their own progenitors—us.

Our world is getting ready to change again. Once again, life, that endless dance of adaptation to the universe and to itself, is about to change all the rules. It's gathering itself for a great leap in intelligence, and the consequences for our species are likely to be extreme.

The Never-Ending Dance

The earth coalesced out of orbiting rocks about 4,500 million years ago. Within a thousand million years, life started. But rather than persisting passively, as rocks do, life changed all the time.

Imagine that you are a single-celled life-form floating in the sea about two thousand million years ago. You haven't much of a brain; actually, as next of kin to a bacterium, you've no brain at all. Luckily, neither does anyone else. So what do you need? First, above all else, when the local food supply runs out you must move. Second, if there are signs in the water pointing to more food, and if you happen to invent senses to detect these signs, you could be better off than others. So being able to distinguish food signs from nonfood signs—a simple piece of information—now becomes more important to you than anything else. Information is the pivot all life's machinery turns around.

Simply detecting things is useless if you cannot also act on them. So you invent a technology to move (muscles won't be invented for a while yet). Of course, once other creatures key in on the advantage of the new technology, they all start moving. Those that don't are outcompeted; they either leave no offspring or become the ancestors of fungi, pond scum, and plants. When everyone can move, simply being mobile is no longer enough. Another thousand million years go by.

You're now a wormlike ribbon of cells and still live in the sea. By now it's roughly seven hundred million years ago, and sex has been invented. Then things really get going. The diversity of living things simply explodes as the new technology for mixing and matching genes leads to rapid change.

Then a terrible calamity happens. Perhaps an asteroid hits, perhaps the earth wobbles, perhaps the sun hiccups. Whatever the reason, many die. But the few survivors don't care, for now the almost-empty ecosystems are fertile ground for a new onslaught. Life explodes everywhere. It's now only five hundred million years ago and you're a kind of bottom-feeding jawless fish. Without knowing it, or planning it, you're slowly inventing the biological technology that will help you, in 140 million years, become an amphibian.

So the long eons march on, while living things, on a long road to nowhere, never giving a thought to the future, occupy themselves (as biologists say) with the four Fs: fighting, fleeing, feeding, and—er—reproduction. Each eon is punctated by a new technology or a terrible calamity resulting in a brief flurry of activity and confusion; the survivors either incorporate a new technology to handle the current crisis or branch off to fill other niches.

Every time evolution throws up a new technology for living life the same thing happens: the eternal arms race starts all over again, with more ammunition or in a different arena. Each arena—each way of living life—becomes a new ecological niche as life-forms compete to live in it. Despite the recurring catastrophes, the eternal competition that is life keeps going—always different, always the same. The names change, but the dance stays the same. That's life at its most elemental: competition and revolution, technology and strife, change and adaptation, chance and necessity.

A Mind of One's Own

The brain started as a simple information manipulator to help the body cope with rapid change. A thousand million years ago, when our ancestors were ribbons of cells living in the sea, life was simple. In those days, your earliest senses—taste, smell, touch—directly affected your muscles. A bad taste? Move away. A good taste? Move closer.

Then everyone in your ecological niche got the same technology. (You had outcompeted all those who didn't.) So to meet the new competition you enhanced your ability to survive by putting nerve cells in the loop. At first, this loop was simply a faster way to transmit messages within your body—to sense danger and contract distant muscles, detect food and engulf it with other muscles. Soon, however, the nerve cells grew numerous enough to start modeling the world, trying to predict events. Then everyone else in your niche did the same thing. The ones that didn't were eaten by those who did. He who hesitates is lunch.

Of course, the life-forms outside your particular niche didn't necessarily have to change, because they didn't compete with you directly. That's why plants, for instance, didn't develop a nervous system and still need only hormones to control their bodies. On the other hand, all many-celled animals except sponges use both hormones and nerves. Lacking nerves, plants are so slow we don't even think of them as moving at all.

Once your direct competitors also had nerve cells, things quickly became much more complicated. You had to recognize not only food cues but also danger cues. Is the water getting too salty near by? Is it getting too hot? Too cold? Is that nearby life-form trying to make you lunch? Would it make a nutritious snack? Is it a potential mate? As you grew more complex, so too did the questions you needed to answer to survive. Meanwhile, you also had to learn to respond more and more rapidly. Here today, lunch tomorrow.

As things got faster and more complex through the never-ending competition with others and with the universe, you needed to have more and ever more nerve cells between your senses and your muscles—you needed clumps of nerves to manipulate information and decide what to do next. As the complexity increased, these clumps began to use memory—a shorthand snapshot of what happened in the past—as a guide to the future.

So your brain tries to predict the future and guide your body to find good tastes and avoid bad ones. Eventually it could do this many times faster than the genes that created it, just as computers now make decisions many times faster than the people who made them. Munch or be munched.

Identifying the Pivot

As soon as one life-form's behavior grows more sophisticated and less instinctive, its competitors' must too. Change or die—those are the usual options. Every living thing works desperately to adapt to changes in its environment and in every other living thing.

That lesson is nowhere more obvious than in the special care our body gives our brain. Our brain is less than a fiftieth of our body mass, yet it uses a fifth of our food and a fifth of our blood. Roughly half our genes are needed to build it and another third to maintain it. The brain is the only organ it would be pointless to transplant and the only one that is totally encased in bone (although the womb comes close). It's also the only organ with a special cleaning system and a dedicated blood supply. A hard blow to the head knocks us out; a blow elsewhere usually just causes pain. We faint—and so become helpless—before letting oxygen starvation harm our brain. We spend a third of our lives in sleep, defenseless. Why?

Our brain is where we make information tangible. Anything we can touch, we can manipulate somehow, perhaps to our advantage. By sticking nerve cells between our senses and our responses a thousand million years ago, we gave ourselves the ability to change our responses to environmental change. And, because nerve cells use only a little energy, it's fast and cheap to adjust. In effect, our brains amplify tiny energy changes into big behavioral changes—and adapting to change is what life is all about. All life on earth constitutes a gigantic adaptive machine, adapting itself both to the vagaries of the universe and to itself. A snake eating its own tail, life eats itself to produce itself.

Over the eons, our ever-larger brain gave us an ever-bigger bag of tricks: more and yet more-varied responses for finer and yet finer distinctions. Every increase led to more sophisticated behavior. Each new behavior is like a new organ; although intangible, it is as real as a kidney. Adding a new behavior is cheaper and faster than creating a new organ, just as reprogramming a computer is cheaper and faster than building a new machine. That's why learning new things, new ways of being and doing, is so important.

So, having a brain is supremely important because tinkering with it can be far faster and easier than tinkering with other organs. And, because tiny changes in it can result in major changes in behavior, any brain change is far more important than any other physical change.

The competition to produce better, cheaper, and more accurate high-speed prediction and control still exists today. It's so important that eventually we, the most complex brains on the planet, have produced external brains to help us handle the insane growth rate needed. We call these external brains *computers.* Just as the presence of the brain radically changed the role of its creators, the genes, so too will the presence of computers radically change the role of their creators, us. Change the brain and you change the world.

Into the Exponential

If we compress all the time since the beginning of the universe into a single year, one day corresponds to about 45 million years and one second is about 521 years. In that calendar, the earth didn't even form until the middle of September, the first life didn't show up until early in October, and the first many-celled life-form only evolved in mid-December. Then, just two days later, the first fish appeared and, two days after that, the first insects. The next day came the first amphibians; the next, the first trees; the next—Christmas Eve—the first dinosaurs.

Three days later, the dinosaurs had vanished. By then their only remaining descendants, the birds, had shown up and mammals were on the rise. The primates showed up the next day. Two days later, December 31, about an hour before New Year's Day, the first protomodern humans showed up. A little over a minute to twelve, the first modern humans appeared—around thirty-five thousand years ago. At half a minute to twelve, the last ice age ended and humans scrambled out of the cave. In a few seconds they started farming and building cities.

At one second before midnight, they began printing books, having a renaissance, and making science a serious subject of study. With half a second to go, they staged an industrial revolution and then, in the last tenth of a second—in just fifty short revolutions of the planet around its sun—these hairless, two-legged, big-brained primates doubled their population, quadrupled their goods and services, tamed the atom, artificially extended their brains, reshaped their planet, jumped beyond it, and started rearranging their genes.

It's hard to think of such a panorama without feeling that it's leading up to something—something big. But, of course, as far as we know, it isn't. Fifty million years from now another species may look back and not even bother to mention us in a similar panorama. From their perspective, everything will have been leading up to *them*. Still, it does seem reasonable to guess that living systems will continue to get roughly more complex.

Of course, that doesn't hold everywhere and for all time. The dinosaurs, for instance, could conceivably have evolved into tool-using, space-traveling, nuclear-bomb-dropping animals just as we did. But something happened, and now they're gone. We too could soon be gone. A few million years here or there doesn't much matter to old mother earth. There's always the chimpanzees—or the dolphins—or the mice.

Yet, unless we're hit by a passing asteroid, manage to do ourselves in, or—ignominy of ignominies—unless our artificial children replace us, it seems likely that we'll continue on an upward spiral of more and yet more complexity. After all, we're still running in the same old evolutionary race.

Boarding the Darwinian Bus

Climbing the ladder of increasing complexity is not a recent thing. For example, the first trees were probably much shorter than the ones we have today. As soon as one tree species in some ecological niche chanced to grow just a bit taller, it forced all other trees in that niche to grow taller, too. Failing to do so meant living in the shadow of the taller trees and so getting less sunlight. As in a self-reinforcing arms race, all the trees grew as tall as was possible, given the limits on incident sunlight, ground minerals, and hydrostatic pressure.

Similarly, once upon a time plants probably couldn't defend themselves against the insects that ate their leaves. Then one plant's cells began, by chance, to produce an insecticide—a chemical that kills insects. Some insects then retaliated by developing immunity to it, and the plants developed new insecticides—which some insects also adapted to. In an ever-escalating cycle, the process continued, until today broccoli, for example, consists about 10 percent by dry weight of natural insecticides.

If at a football match everyone in front of you stands up to see better, you must stand up too—if you want to see the game. Then, if all the others stand on their seats, so must you. Like standing on seats, escalation can stop, of course; but anyone who tarries too long on the highway of life is likely to get run over. Zebras start running a little faster to escape lions, so lions have to run a little faster to catch zebras. But then zebras have to run even faster, because while the lion is running for its lunch, the zebra is running for its life.

We're caught up in the same old arms races today, although—because we've managed to outcompete most of our usual predators and parasites—we mainly race with ourselves. Today, internal forces are driving us to augment our intelligence and create machine intellects. Any massive intellect, however, will eventually gain the power of life and death over everyone it serves. So what life-forms, you might ask, would be stupid enough to engineer their own demise? We would.

We will do so, in all probability, because of military and commercial escalations. For example, most major cities are between six and ten minutes away from a nuclear submarine attack. We can't juggle all the options in that short a time, so we're shunting ever more of the routine decision making to ever-faster computers.

Businesses too need more capable and more autonomous machines for exploration, control and prediction, and increased production. Few of us are willing to work at dirty, dull, demanding, or dangerous jobs. How would you like to be a sewage disposal engineer at a nuclear waste site? The problem is that once we make machines competent enough to do these kinds of jobs, they'll also be competent enough to collect our garbage, mow our lawns, clean our houses, and guard our property. Most of us who once did those things will then be unemployed—outcompeted by machines.

Once one business develops a new and complex computer system to stay one step ahead—whether in a bank, a chemical refinery, or a power plant—competitors will have to get one as well. If they don't, they go under. Lunch or be lunch.

Every step up the ladder of rising complexity forces us to take the next step up. And each step leads to unforeseen new steps. For example, imagine how happy disabled people would be with a legged wheelchair that can handle all terrains controlled by their thoughts alone. The military and industrial uses of such a device, however, would be even more dramatic. If ever such a breakthrough occurs, more than the lives of the handicapped would change.

Imagine, for example, how happy a military service would be to get a small, cheap, legged robot that can crawl up a drainpipe or down a shoreline carrying a listening device—or an explosive. Only a few decades later those robots would be so cheap, effective, and widely available that civilians could use them to spy, steal, or kill. Then, in the usual escalating arms race, security services would have to use them for defense. Then criminals would use better ones to regain their edge. And so it goes. Each escalation pushes us into the next escalation.

Eventually, we may end up with machines so complex that none of us will truly understand them. That point may not come in the unimaginably far future, either. We may reach it in forty years. Just twenty years separate Yury Gagarin's *Vostok I* flight from the first launch of the shuttle *Columbia*. Just thirty-five years separate the discovery of our genetic structure from the first patented artificial animal. Just two generations watched the *Kitty Hawk* evolve into stealth bombers, *Sputniks* into *Voyagers*, waterwheels into nuclear power plants, and fireworks into hydrogen bombs. Our technology is now melting, flowing, and reforming into new and alien shapes, faster and faster, so fast that prediction has become nearly impossible.

Careening Between Pollyanna and Cassandra

Imagine that you're reading a book through a tiny window. At any one time you can see only a few letters. Could you make sense of the book? We live all our lives in that window. In a tiny warm eyeblink we must learn all that our culture requires of us and somehow fashion a life. Just when we think we understand enough of the universe to live well, the eye blinks and our life ends.

Slow technological change made our lives easier in the past because we only had to learn what our ancestors had already deduced. We didn't have to figure out anything new for ourselves; we merely had to learn what our parents and community thought it best for us to know. In those days, it was silly to ask what the future would be like. The future would be just like the past.

Today though, asking about the future is like asking a blind man to describe the shape of a flame. Having never seen it and having no way to touch it, he can only resort to reason and inference. No matter how much information he gathers, he can never succeed because the flame is always dancing in the winds of change.

Predicting even our near future is already impossible. Too many things can change, and change radically. Even when they don't, they may be used in totally unforeseen ways. Suppose, for instance, that you and your friends are on a trek and decide to explore a cave. You are carrying guns to protect yourselves against wild animals but lose all the rest of your equipment in a cave-in. You're trapped underground with only your guns. Do you give up and die?

Some would. But if it happens to enough people, eventually some group will try using their guns in new ways. They might, for example, throw away the guns, dismantle the bullets, and use the gunpowder to blow up the rubble blocking their exit.

That's exactly how evolution works. All living things have various abilities that are finely honed for survival in their normal environment. In a new environment, however, many of those abilities are useless. Sometimes, though, a few of them can be used in slightly different ways and whoever adapts them to the new purpose lives another day. So if explorers habitually frequent caves, they might eventually carry bags of gunpowder instead of guns. Life is make do or die.

Evolution is a continual dance of new technologies for living life, always with too few resources and always with only one penalty for failure—extinction. Each such new technology is largely a chance collection of different abilities—slightly taller trees, slightly deadlier plants, slightly faster zebras—generated by eons of tiny random changes. Technology isn't, therefore, some new thing recently invented by mad scientists to make our lives miserable; it's woven into our bones and blood, into our very existence. Life itself is the ultimate technology.

Technology Is Our Iron Lung

Technology drives our civilization. Yet, today, as its rate of change accelerates, more and more of us in advanced countries seem to be disenchanted with it. Many of us leave our comfortable, well-heated homes only to burn enormous quantities of fossil fuels to toddle down to our clean and insanely well-stocked grocery stores, credit cards clutched firmly in hand. Too many of us are far, too far away from disappointing harvests, aching backs, and sunstroke to properly appreciate the rigors of the peasant life some seem to long for so dearly.

In advanced nations it's even becoming quite fashionable to reject technology entirely and long for some mythical past. But romantic fantasies of the past focus can only on the tiny historical minority who had power to make a decent life for themselves. Because for the majority of our ancestors, life was pretty horrid and only technology saved us from it.

For most of western history, most people were poor, ignorant, dirty, and diseased. There was no such thing as hygiene. Even the grandest cities stank. Streets were open sewers. Waterways were polluted, and plagues were as natural and inevitable as the sunrise. Doctors, only recently risen from the ranks of butchers and barbers, killed and maimed as many patients as they cured. The cure for a hemorrhage was to bleed the patient. Most people's teeth rotted and fell out in their late twenties.

For eons, we killed our surplus infants as often as we killed our implacable enemies. Famine wasn't rare—it was a way of life. People were hanged for stealing a loaf of bread. One in ten women died in childbirth. Women were possessions. Children didn't go to school; they went to work. Even after African slavery was abolished, effective European slavery continued for the working poor until the late nineteenth century. And for the poor—which used to be almost all of us—there was no redress in the law, because going to court often meant being tortured by the judges.

In sum, until our very recent past, families were big, infant death rates were high, the middle class was tiny, justice was a joke, illiteracy was the norm, starvation was common, and medicine was a bottomless pit of ignorance.

Science and engineering changed all that. To give just one example, lifespans in advanced countries rose more in this century than they had in all the previous five hundred centuries of recorded history. At the beginning of this century most of us died in our forties, just as we had since time immemorial. Now, at century's end, those of us lucky enough to be born in advanced countries die in our seventies.

Since the first human hand snatched up a fiery branch accidentally created by lightning or volcanoes, we've been continually amazed by the possible. Fifty years ago we couldn't imagine changing our genes. A hundred years ago we only dreamed of flying through the air. Three hundred years ago we couldn't conceive of invisible microbes. We utterly failed to imagine all sorts of things we now take completely for granted: the Pill, electricity, X-rays, printing presses, antibiotics, aspirin, plastics, artificial fertilizers, moon shots, computers, biotechnology. All caught us by surprise.

Science gives us that cornucopia because we use it to ask the universe what things are really like, not what we would like them to be. Often the answers we get are startling, and they change our very lives. Yes, it *is* possible to generate power from certain rocks. Yes, it *is* possible to compute with certain other rocks. Yes, we *can* bioengineer ourselves. Science and engineering are so productive, so pivotal to improving our lives because the universe has a far better imagination than we ever will.

The arts and the humanities give our lives meaning, but science and engineering give us the means to live. Engineering supplies new tools for living, while science offers us new ways of thinking. The enemy has always been ignorance, not science and engineering. Technology has serious problems, yes. But the alternatives are worse.

The New Millennium

Largely because of cheap, powerful computers, we're now entering a period of very rapid change. If, as they say in Texas, education follows the second kick of the mule, today's infotechnology is only the first kick. Tomorrow's biotechnology will be the second one. When that second wave finally passes it will leave us, like beached fish, gasping for air on a new and alien shore.

We can already inject foreign genes into most living things: firefly genes into tobacco plants; tobacco genes into pigs; pig genes into mice; mice genes into sheep; sheep genes into cows; and human genes into cows, mice, sheep, pigs, tobacco plants—and the reverse. The ability to do this has consequences—enormous consequences.

Today, blood is in short supply. Few people donate it and what they do give has to be filtered and tested, making it expensive. Despite the testing, of the 3.5 million Americans who get blood transfusions each year, over a thousand contract hepatitis. Who wouldn't pay to get cheap, untainted, human blood from pigs or sheep or cows if it meant the difference between life and death? Well, we can do it—not tomorrow, not next year—today.

Organ donors are also in short supply. People don't sign their donor cards, accidents happen far from hospitals, relatives are too busy grieving, and medical teams are reluctant to ask. Few of the donated organs can be used anyway: they may be the wrong size; the blood type may be wrong; there may be complications. Consequently, even in advanced countries, fewer than one in a hundred people who will die without donated organs get them. And one in five of those who do may still die, because of tissue rejection. Who wouldn't pay for a personal transgenic pig to carry copies of all important organs so that if one fails—a heart, a lung, a kidney—it can be replaced? Such a replacement couldn't be the wrong size, nor would it be rejected, because it's as identical to your own organ as if it grew in your body.

In 1990s America alone, health care costs consume nine hundred thousand million dollars annually—that's almost a seventh of the entire economy. It is estimated that Americans spend another four thousand million dollars a year on antiaging cosmetics. Further, every year, roughly a million Americans spend large sums on cosmetic surgery—not to mention the enormous amounts spent on dieting and exercise.

Our body is at its peak at about twelve years old. It remains supple and efficient for six long years, then, at about eighteen, begins to die. Beyond eighteen, our brain cells die in ever-increasing numbers, our teeth fall out, our immune systems lose flexibility, our muscles and bones thin and weaken, and our skins thicken and roughen. By twenty-one, the hand of death is already heavy upon us, and its weight rises with the passing years. By thirty, all our bodies have to look forward to is disability, disease, and dependence.

Evolution has shaped us to fit that cycle because, from the point of view of our genes, we're only useful until we pass them on—which can be as early as twelve or thirteen. Once we can reproduce, we're only useful to our species for as long as it takes to raise a few children to reproductive age—maybe another fifteen years. Beyond, say, thirty, evolution hasn't worked out the remaining kinks in our design. It would have been a waste of effort. So it didn't.

To us, that seems like a terrible waste, but look at it from our genes' point of view. Each of us alive today is descended from a unbroken chain of people, each of whom were once young but few of whom were ever old. Over millions of years, evolution has had a chance to weed out most of the genes that lead to early death; so these genes were rarely passed on. In all that time, however, it has had little chance to weed out genes that kill us later in life. But that doesn't mean we can't do something about it.

For example, there's a rare disease called *progeria* that hastens aging. Twelve-year-olds with progeria look like tiny seventy-year-olds, and they soon die from all the ailments seventy-year-olds die from. The disease results from one change in a single gene. So it may be that many of our old age ailments can be traced to similarly flawed genes. If we can find them and fix them, the ailments should vanish.

Now try to imagine a world, perhaps fifty years hence, where we can get baby-soft skin from body paint, new teeth from chewing gum, strong and flexible bones from milk, superhuman muscles from an injection, greater intelligence from an inhaler, higher sex drive from a pill, better memory from hair spray. That kind of technology could force major changes in how we think about the world and ourselves. For example, what might Muslims and Jews do if we insert pig genes into tomatoes to make them riper or tastier or longer lasting? Does their religion let them eat such a hybrid? Can Roman Catholics eat a fish during Lent if half the fish's genes come from cows? Can Buddhists eat corn rebuilt with human genes? When these religions began such questions were meaningless.

What will we do if we find genes that retard or reverse aging? Or genes that retard or reverse cancers? Or genes to make childbirth painless and danger-free? Or ways to permanently change our height, weight, sex, race, intelligence, memory, patience, skill? Who will let themselves be engineered to carry those genes? Will we let our governments determine the genetic structure of our children? The questions go on and on, one leading to another. For instance, is it okay for the government to check an airline pilot's genes for propensity to heart attack? If so, is it also okay for an employer to check your genes for propensity to Alzheimer's disease?

Biotechnology isn't something we can safely leave to our next generation to think about; we're already using it. So the question isn't when we'll start, but when do we stop? To ask the question is to answer it: We won't stop. Just as we aren't stopping today with computers. We've always left the consequences to our descendants. Why should we change now?

Of course, there will be slowdowns and setbacks here and there as this country or that religion decides that one change or another is too dangerous, too immoral, or just plain too strange. But the incentives to change are simply too high to turn back now. Since its beginnings 3,500 million years ago, life has always been make do or die. Here today, lunch tomorrow.

Where the Wild Things Are

Today, we live all our lives in the cage of our flesh, giving thought to it and its immense complexity only when it fails us. Thanks to technology we're now beginning to ask questions about that cage that our ancestors never dreamt of. We're beginning to dream impossible dreams. Is sleeping away a third of our already oh-so-short lives really necessary? Could we enhance our brain artificially? Could we transplant our brain? Is death really necessary?

For an engineer, every object is just a set of atoms arranged in some particular way. Since all atoms of one type are the same and only atomic organization matters, we could duplicate any molecule by building another one out of the same set of atoms. If we can copy a molecule, why couldn't we, one day, copy a person? If we could do that, why couldn't we postpone death forever? If we could manipulate single atoms cheaply, and if we had cheap computing power to track all the details, we could do almost anything. Nor is this complete fantasy.

In our very far future—fifty to a hundred years from now perhaps—we'll probably have horrifying wars, serious genetic accidents, massive job losses, astronomical differences between haves and have-nots, and major computer-related catastrophes. But we may also have people who continually refashion their bodies as easily as we reshape plastics. That future may be a world of augmented people, heavily armed guard dogs, and men who carry babies to term. Perhaps there'll be cows bioengineered to give birth to supercomputers, medicine in clothes, riot-control joy mists, antimugger fear-inducing aerosols, people who live for centuries, and houses smarter than monkeys. Some of us may have accelerated childhoods and painless childbirths. We might get medical care from a machine, be born with augmented senses and reinforced bone and muscle, and be able to build robot exoskeletons for ourselves and our aging pets. We may even have the artificial equivalent of telepathy and telekinesis.

On the other hand, even if the fountain of youth is found, perhaps only the rich will drink from it. What might it be like to live in a world where the vast mass of us seem like ignorant and retarded peons compared to the elite? What will become of the social structures we've built up so painstakingly by assuming equality for all? What kind of laws will we have, and how will we enforce them? How will we live?

We're now in a curious position: Most facts we learn in childhood can be obsolete by the time we become adults. What will it be like when basic assumptions about how the world works and what is technologically possible change every five years? Given the present stupendous rate of technological change, it's almost senseless to extrapolate twenty years into the future, far less fifty. Provided we don't destroy ourselves first, the world of fifty years from now may be as different from today's as our era is from the Stone Age. The world of a hundred years from now may as well be another planet. And the human race may still call itself human, but it may as well be a new species.

The End of the Beginning

Technology isn't just computers and space ships, video games and cellular phones, atomic microscopes and nuclear weapons. It's plows and needles, paper and pottery, buttons and cloth, light bulbs and plastic bags, shoes and flush toilets. These things don't grow on trees. We made them.

Of these artifacts, information manipulators—devices that can sense and respond to changes in their environment and that can follow long and intricate programs of behavior—are the most recent, the most unusual, and perhaps the most feared. Still, since at least the nineteenth century, various technological innovations have been worming their way into every nook and cranny of our society. They now run our civilization for us. Without our numerous artificial helpers we'd often be too slow, too few, too expensive. The catch is that by making and using technology we've made it indispensable. With it, we've quintupled our population in under two hundred years—an unheard-of growth rate for our species. Without it, four of every five of us could die.

Like all technologies, computers have consequences. But because our brain is the source of all our technology, and because computers let our brains think better and faster and cheaper, they have consequences all out of proportion to those of most technologies. Change the brain and you change the world. For example, today's computers are helping us create biotechnology by letting biologists manage the complexity of their task on all levels—from understanding molecular images to organizing drug schedules; from searching for similar genes to handling inventory; from extracting genes to controlling the centrifuge. Take computers away and the pace of biotechnological change would slow to a crawl. As would the pace of change in every other technical field.

Their most important consequence, however, is that the better they are, the better their next generation will be. If each future computer generation midwifes changes as massive as biotechnology, and each generation arrives in half the time of the last one, what might our world become in thirty years? Today, the new thing is the computer. Tomorrow, it will be biotechnology. The day after, it may be nuclear fusion. Next week, machine intellects. The week after, perhaps anything we want made to order, atom by atom. Brave new devices for a brave new world introduced at brave new speeds.

The scientist and the engineer give us those things, but how we use them is up to us. It sounds wonderful to say that our governments should protect our way of life so that we can keep doing the same old things in the same old ways our ancestors did. The same line of thought goes on to say that scientists should stop discovering, inventors should stop inventing, and engineers should stop building. We're comfortable enough as it is. Leave well enough alone.

Well, it sounds wonderful, but it isn't going to happen. Not just because we don't really want it to, but because three-quarters of the world's population isn't anywhere near as well fed, well off, well educated, or well armed as we are. And they definitely want to be. The sooner we accept that, the better are our chances of surviving the future we're so busy crafting for ourselves.

We can no longer expect to learn to live a certain way, then just sit back and vegetate for the rest of our lives. That way lies disaster. Our sole advantage over today's machines is that we can still change faster than they can. We can adapt, and they still can't. So let's use our wonderful adaptability to keep one step ahead of their inevitable encroachment. And let's pray for guidance when comes the inevitable day that they can adapt too; for then, we'll have no advantage at all.

Let There Be Light

Imagine that you're a disembodied eye floating in the darkness of space, surrounded by a vast ocean of gas and exceedingly fine dust. Now watch those two dust grains nearest you: they're ever so slowly moving toward each other. Gravitational attraction is weak when the masses are small, but it's always present no matter the scale. Eventually, some of the nearby grains get close enough to orbit each other in a loose ball. Once that happens, the local concentration of mass becomes a gravitational magnet for nearby grains. More and yet more grains start drifting toward the tiny ball. As they start to orbit it, the local force of gravity grows larger and larger.

As the local attraction increases, more grains fall in, until the local force grows so strong that other balls of loosely coupled grains start drifting into the growing mass. It's like watching a slow-motion video of water whirlpooling down a drain. The bigger the ball, the bigger it gets. The bigger it gets, the faster it grows. What was once two tiny grains circling each other simply grows and grows until it's as big as a rock, then a planet, then tens of thousands of planets. As the huge ball compresses under the force of gravity, its core temperature eventually heats up to a few hundred thousand degrees and it becomes a superhot plasma.

More quickly now, the plasma gets hotter and hotter, and denser and denser, until the core temperature passes a few million degrees. The enormous heat and gravity then fuse the dense plasma in its very heart, and a thermonuclear fusion reaction begins. A lot of exciting things now happen very quickly. The fusion in the core radiates enormous quantities of energy. The high temperatures and the continuing fusion encourage more and yet more plasma to fuse in a nuclear chain reaction. The enormous outward pressures from the massive amounts of energy being released prevent the outer shell from falling into the inner core and the huge ball stops shrinking. The fusion reaction is now self-sustaining. The ball of once-dead matter ignites and becomes a sun.

We seem headed for our own starbirth, drawn to it just as inexorably as those grains merge, accelerating toward it just as surely as the merging accelerates. The attraction is massive, relentless, unstoppable. When our starbirth comes, some of us will no longer be truly human; and things we now call machines will no longer truly be machines.

Trembling on the Brink

Security is mostly a superstition. It does not exist in nature, nor do the children of men as a whole experience it. Avoiding danger is no safer in the long run than outright exposure. Life is either a daring adventure, or nothing.
Helen Keller, *Let Us Have Faith*

Every decision requires a choice, for with limited resources we can't pursue all options forever. Yet making each choice locks us into a certain path, and, as we make more and more choices, our path becomes more and more definite and less and less like any other path we could have taken. Because we can't foresee all possibilities, the path we end up walking often isn't the path we intended to walk when we took our first faltering step.

In our time, that first step was the invention of the computer. Although few realized it at the time, the computer is an undifferentiated information manipulator. So we can use it to extend our intelligence, just as we use books to extend our memory. As it isn't specific to any one task, we can use it for everything. And that fact changes everything.

We're good at some things and bad at others. Doing long calculations and weaving intricate silk patterns are two things we're bad at. So we invented a device that can do easily what we find hard. Later we found many other uses for our new device, and these new uses started forcing us to leave our old, comfortable occupations and find other jobs. That ejection from traditional ways of life also freed us to do things we enjoy and are good at. Unfortunately, that same machine may one day become good at the things we're good at. What will we do then?

Our society is now undergoing enormous technological changes. The computer is at the center of them all—if not as a direct cause then as an irreplaceable helper. Computers are everywhere, and they're moving deeper into the woodwork and farther from our sight all the time. In time, they'll slip inside us and some of us will become something new— something alien.

Like the steam engine, like the dynamo, like the telephone, the computer is an accelerator. It accelerates every process in our society. By letting us think better, faster, and cheaper it's hurled us into an era of technological change undreamt of just fifty years ago. But that pace is hard for us to manage because not long ago our lives moved to our pulse beat. Nothing could happen much faster than that because nothing moved to our command unless we moved it first. Now that we've deployed millions of decision-making machines to do things for us, however, our lives move to a metronome's beat. And someone, somewhere, is always cranking up the tempo—and pumping up the volume.

Our technology changes far faster than our biology can. Our biology hasn't changed for thirty-five thousand years. Evolution has spent millions of years shaping us to be very good at coping with the world as it was up to, say, two hundred years ago when automation first began to seriously bleed into our societies. Today though, our technology is far beyond fire and stone and water, and we're continually playing catch up with it—never quite in time with the beat, never quite ready for the next set of dance steps it requires of us.

That's been true for at least the last two centuries, ever since the industrial revolution. What's special today is that because of the computer, an undifferentiated intelligence amplifier, our technology has nearly reached critical mass and is now juggernauting us around the dance floor at such a pace that we may never again be able to stop and catch our breath. Now prisoners of the dance, we're moths irresistibly attracted to the flame of technology. Prometheus, disguised as a scientist, has given us that flame. But fire also burns.

In the long human haul that is our life, we're now forced to gouge out our hearts and slowly murder our old selves in a never-ending search for new ways to live in this, the newest empire our own minds have made for us. Enmeshed deep within the mechanism and crying for simpler times, we seem able only to muddle along in idle despair, cringing at the coming of our own new millennium.

But while change is painful, it's necessary too. Without it, without turbulence and decay and unity in variety, there's no rebirth, and life stagnates and dies. We fear change, but perhaps we should fear lack of change more, because the ultimate lack of change is death. It's change itself that brings meaning and richness to life. How peaceful would be our lives without all this constant change; how peaceful, how tranquil—and how dull.

A strange new world is coming, and coming fast, partly brought into being by a strange new machine. For various short-term and inescapable reasons of our own we're rapidly creating an empire of the mind, and now we must find some way to dwell in it. Our future is filled with enormous danger, yes, but it's also filled with unimaginably exciting possibility. Everywhere, life seems to be gathering itself for a great leap forward. After millions of years of slow germination, we're rapidly beginning to flower.

We, all of us, are part of the most thrilling adventure ever unleashed on planet earth. Instead of looking backward in anger and fear, let's look forward to the next dance step in the adventure we're crafting for ourselves. A century or so from now, the earth may simply be the home world of a species rich and strange, a fiercely new and amazingly interesting species—transhumanity. The human adventure is just beginning.

Let's dance.

My Thanks

I thank my agent, Laura Fillmore, and my publisher, Harry Brad-ford Stanton, for a great job and true belief. I also thank the follow-ing people who read and commented on various drafts and helped in numerous ways: Angela Allen, Ricardo Baeza-Yates, David Bartlett, Philip Bradford, Phillips Bradford, Randy Bramley, Jeanette Calvert-Coffren, Michael Chui, Eliana Colunga-Leal, James Conley, Caroline Countryman, Joe Culberson, Cornelia Davis, Susan Doherty, Ruth Eberle, Vladimir Estivil-Castro, Julia Fisher, Janet Foster, Lisa Free-man, Juliet Frey, Dan Friedman, Adrian German, Glenn Goldstein, Jodi Graham, Brian Gygi, Andy Hanson, Steve Hayman, Manoj Jain, Michael Jensen, Steve Johnson, Elizabeth Jones, Terry Jones, Rick Kaz-man, Jim Kling, Kate Ksiazek, Lorrie LeJeune, Yue-Herng Lin, Tom Lipscomb, Sushil Louis, Jim Marshall, Gary McGraw, Suzanne Menzel, Steve Miale, Jon Mills, Sue O'Rourke, Al Paeth, Bob Port, Darrell Ray-mond, Jean Reese, John Rehling, David Rosenblueth, David Roth, Doug Rousch, Lorilee Sadler, Airlie Sattler, Elizabeth Schneider, Pete Shirley, Raja Sooriamurthi, Bruce Spatz, Mike Sullivan, Shankar Swamy, Ellen Tate, Raja Thiagarajan, Sr, Mark Tilden, Gina Torrett, Venkataraman Vaidayanathan, Dirk Van Gucht, Marc VanHeyningen, Oscar Waddell, Pei Wang, David Wise, and Derick Wood.

For help far above and well beyond the call of duty I particularly thank: Judi Copler, Mert Cramer, Bill Dueber, Julie England, Jerry Fenner, Beth Freeman, Nola Hague, Merav Harris, Chris Haynes, Mary Mahoney-Robson, Karen Miller, Leslie Ortquist, Jacqui Pulliam,

John Pulliam, Malcolm Rawlins, Stephen Rawlins, Steve Ryner, Derek Smith, Raja Thiagarajan, Laura Watkins, Dedaimia Whitney, and Laura Wright. Thank you all.

Gregory J. E. Rawlins, rawlins@cs.indiana.edu
http://www.cs.indiana.edu/admin/faclist/rawlins.html

Index